Tributes
Volume 22

Modestly Radical or Radically Modest
Festschrift for Jean Paul Van Bendegem on the Occasion of his 60th Birthday

Volume 12
Dialectics, Dialogue and Argumentation. An Examination of Douglas Walton's Theories of Reasoning and Argument
Chris Reed and Christoher W. Tindale, eds.

Volume 13
Proofs, Categories and Computations. Essays in Honour of Grigori Mints
Solomon Feferman and Wilfried Sieg, eds.

Volume 14
Construction. Festschrift for Gerhard Heinzmann
Solomon Feferman and Wilfried Sieg, eds.

Volume 15
Hues of Philosophy. Essays in Memory of Ruth Manor
Anat Biletzki, ed.

Volume 16
Knowing, Reasoning and Acting. Essays in Honour of Hector J. Levesque.
Gerhard Lakemeyer and Sheila A. McIlraith, eds.

Volume 17
Logic without Frontiers. Festschrift for Walter Alexandre Carnielli on the occasion of his 60th birthday
Jean-Yves Beziau and Marcelo Esteban Coniglio, eds.

Volume 18
Insolubles and Consequences. Essays in Honour of Stephen Read.
Catarina Dutilh Novaes and Ole Thomassen Hjortland, eds.

Volume 19
From Quantification to Conversation. Festschrift for Robin Cooper on the occasion of his 65th birthday
Staffan Larsson and Lars Borin, eds

Volume 20
The Goals of Cognition. Essays in Honour of Cristiano Castelfranchi
Fabio Paglieri, Luca Tummolini, Rino Falcone and Maria Miceli, eds.

Volume 21
From Knowledge Representation to Argumentation in AI, Law and Policy Making. A Festschrift in Honour of Trevor Bench-Capon on the Occasion of his 60th Birthday
Katie Atkinson, Henry Prakken, and Adam Wyner, eds.

Volume 22
Modestly Radical or Radically Modest. Festschrift for Jean Paul Van Bendegem on the Occasion of his 60th Birthday
Patric Allo and Bart Van Kerkhove, eds.

Tributes Series Editor
Dov Gabbay   dov.gabbay@kcl.a

Modestly Radical or Radically Modest

Festschrift for Jean Paul Van Bendegem
on the Occasion of his 60th Birthday

edited by

Patrick Allo

and

Bart Van Kerkhove

© Individual author and College Publications 2014.
All rights reserved.

ISBN 978-1-84890-070-7

College Publications
Scientific Director: Dov Gabbay
Managing Director: Jane Spurr

http://www.collegepublications.co.uk

Cover by Laraine Welch
Cover photograph by Patrick Allo
Printed by Lightning Source, Milton Keynes, UK

All rights reserved. No part of this publication may be reproduced, stored in a retrieval system or transmitted in any form, or by any means, electronic, mechanical, photocopying, recording or otherwise without prior permission, in writing, from the publisher.

Contents

Author Affiliations vii

Editors' Preface ix

1 How Some Infinities Cause Problems in Classical Physical Theories
   David Atkinson and Jeanne Peijnenburg 1

2 The Consistency of Peano Arithmetic. A Defeasible Perspective
   Diderik Batens 11

3 How Mathematicians Convince Each Other, or "The Kingdom of Math is Within You"
   Reuben Hersh 61

4 Why Philosophy and the Humanities Matter a Lot
   Rik Pinxten 97

5 Mathematical Pluralism
   Graham Priest 111

6 On Play-Objects in Dialogical Games. Towards A Dialogical Approach to Constructive Type Theory
   Shahid Rahman, Nicolas Clerbout and Zoe McConaughey . . 127

7 What Can A Sociologist Say About Logic?
   Sal Restivo 155

8 **Mysterianism Revisited: On The Semiotics of Consciousness**
Roger Vergauwen **179**

Author Affiliations

Editors

Patrick Allo, Bart Van Kerkhove

Centre for Logic and Philosophy of Science
Vrije Universiteit Brussel, Belgium
patrick.allo@vub.ac.be, bart.van.kerkhove@vub.ac.be

Contributors

David Atkinson

em., Quantum Gravity Group
University of Groningen, The Netherlands
d.atkinson@rug.nl

Diderik Batens

em., Centre for Logic and Philosophy of Science
Ghent University, Belgium
diderik.batens@ugent.be

Nicolas Clerbout

Sciences, Texts, and Language Research Unit
Université de Lille 3, France
nicolasclerbout@wanadoo.fr

Reuben Hersh
em., Department of Mathematics and Statistics
The University of New Mexico, USA
rhersh@gmail.com

Zoe McConaughey
Sciences, Texts, and Language Research Unit
Université de Lille 3, France
zoe.mcconaughey@etu.univ-lille3.fr

Jeanne Peijnenburg
Theoretical Philosophy Group
University of Groningen, The Netherlands
jeanne.peijnenburg@rug.nl

Rik Pinxten
em., Centre for Intercultural Communication and Interaction
Ghent University, Belgium
rikpinxten@yahoo.com

Graham Priest
Department of Philosophy
Universities of Melbourne, Australia, and St Andrews, Scotland
Graduate Center, City University New York, USA
g.priest@unimelb.edu.au

Shahid Rahman
Sciences, Texts, and Language Research Unit
Université de Lille 3, France
shahid.rahman@univ-lille3.fr

Sal Restivo
em., Department of Science and Technology Studies
Rensselaer Polytechnic Institute, Troy NY, USA
Senior Fellow, University of Ghent, Belgium
salrestivo@hotmail.com

Roger Vergauwen
Centre for Logic and Analytical Philosophy
KU Leuven, Belgium
roger.vergauwen@hiw.kuleuven.be

Editors' Preface

Jean Paul's Paradox *A paradox embodied, rather than discovered by Jean Paul Van Bendegem (b. Ghent, 28 March 1953). Adopting a radical position, which is then only modestly defended, perhaps with a little reluctance or shyness, but never abandoned. This paradox doesn't ask to be resolved, for it almost defines its originator. Yet, it can leave one wondering whether he's modestly radical, or radically modest.*

It is of course no one's personal merit to turn sixty. That is to say, except for staying in good health long enough, it just happens. Be that as it may, not only is it customary to celebrate this particular moment in life of a well respected academic, it is also a highly suitable occasion, for the honored has accomplished most of his or her career while at the same time remaining in full business. Well, not quite, at least in this case. In Belgium, the age of sixty marks the point at which an academic can choose to retire. Although far from actually having done this, nevertheless Jean Paul Van Bendegem at this stage made a bold decision, and after decades of relentless academic commitment went half time at his (and our) home university as of October 2013. This is not the place to address the motives behind this move, well considered as they were, so let us suffice with pointing out the obvious benefits it will have for him in years to come, namely that he can now devote more of his precious, finite time to some of the activities he loves so much and is also terribly good at, as there are: giving popular lectures throughout the country; writing texts of various (and not at all strictly academic) nature; actively participating to cultural and media events; collecting, reading and commenting philosophy and art books, but also loads of novels (those who have had the opportunity of checking out his home office and library will surely appreciate what we are talking about).

The latter interests illustrate why it is risky business to try and put one's finger on the significance of Jean Paul as an intellectual. The consequences for this collection are that some kind of pre-selection was in order. Because a similar compilation of course always retains both a subjective and random character, we have decided to set out with a primary focus on aspects of his scientific work, and invited people from outside Jean Paul's immediate working environment to address themes that along the road have been food for discussion between them. That being said, given these difficulties and the choices we had to make, we are proud to observe that, under the umbrella of humanism in all earthly affairs, this Festschrift does exhibit a substantially representative character after all. Indeed it includes considerations by various of Jean Paul's most important intellectual associates, ranging from some of his earliest colleagues, to local and foreign scholars that have deeply influenced him (and vice versa) in the course of his career. We leave it to the reader to explore and appreciate though, and would like to close this brief editorial introduction with apologies and acknowledgements.

First, an apology, for as Jean Paul Van Bendegem turned sixty in the spring of 2013 already, this tribute volume is clearly overdue. Given Jean Paul's extensive experience with the pitfalls of being an editor, we can't fool him about the causes of such delays. Needless to say, we take full responsibility. Next to that, we would like to express our indebtedness to all authors for their contributions, as well as to our colleague Jonas De Vuyst, for allowing us to make use of a book template previously developed by him, and to the people at College Publications, for their patient confidence. Our final words of gratitude we direct to the celebrated himself, simply for being who he has been and continues being: a humorously understanding, passionately broad-minded, and humbly dedicated chief of staff. So we say in name of all, past and present, at the Brussels Centre for Logic and Philosophy of Science.

December 2013,
Patrick Allo and Bart Van Kerkhove

CHAPTER 1

How Some Infinities Cause Problems in Classical Physical Theories

David Atkinson and Jeanne Peijnenburg

1 Finitism in Classical Mechanics

We all know it: Jean Paul Van Bendegem does not like infinity. In his view, both mathematics and physics can do extremely well without the concept of an actual, or even of a potential infinity. And although he realizes that there is a downside to relegating infinity to the scrapheap of mathematical history, he has always insisted that the advantages outweigh the drawbacks.

Far be it from us to question the feasibility of Jean Paul's program: his ideas contain many valuable insights that should be taken seriously. We do not challenge the consistency of the strict finitist reconstruction of large parts of mathematics; nor have we the temerity to deny that infinity has given rise to many a headache for the physicist. Indeed, we concur with his claim that:

> ... infinities cause all sorts of bizarre problems in the framework of classical mechanics. (Van Bendegem 1992, p. 33)

Yet we will argue that, in classical mechanics, only actual infinities give rise to difficulties, while potential infinities are not only harmless, but in fact quite useful. That explains the title of our paper, which differs by only one word from the title of Van Bendegem's cited 1992 paper, 'How Infinities Cause Problems in Classical Physical Theories'.

One of the headaches of the purveyor of classical mechanics is that determinism can fail if infinities are allowed to creep into the theory. In fact it is claimed on page 35 of the cited paper that determinism in classical mechanics *will* fail if any *one* of the following conditions is violated:

(C1) No forces may be infinite

(C2) No masses may be infinite

(C3) No accelerations may be infinite

(C4) No velocities may be infinite

Interestingly, much of what Jean Paul says in this paper anticipates a completely independent assault on determinism which started in 1996 and has continued to the present day. We mean the assault launched by Pérez Laraudogoitia (1996) and further developed by Alper and Bridger (1998). We propose to describe first Pérez Laraudogoitia's infinitistic model of colliding balls; and then we shall confront it with Van Bendegem's conditions (C1)–(C4). Next we discuss the reaction to this work by Alper and Bridger. The essence of Alper and Bridger's model was already put forward by Van Bendegem (1992); we shall however argue that neither his resolution of the looming paradox, nor that of the later authors, is entirely satisfactory.

2 Zeno balls

In 1996 the Basque philosopher Pérez Laraudogoitia published a paper in *Mind* with the modest title: 'A Beautiful Supertask' (Pérez Laraudogoitia 1996). In it he single-handedly demolished the laws of conservation of energy and momentum, and moreover he breached that bastion of Newtonian mechanics, determinism. We propose to examine his model, and to see which of Van Bendegem's conditions (C1)–(C4) has been contravened.

In general, any system containing a finite number of balls that undergo a finite number of elastic collisions amongst themselves respects the laws of conservation of energy and momentum. However, consider the following idealized system. An infinite number of identical point masses (balls) are placed at the Zeno points $1, \frac{1}{2}, \frac{1}{4}, \frac{1}{8} \ldots$ on a straight line. All the balls are at rest except the first, at 1, which moves with constant speed towards the second, at $\frac{1}{2}$ (see Figure 1). After elastic collision the first ball comes to rest, passing all its kinetic energy on to the second ball, which soon collides with the third ball, which acquires all the energy, and so on *ad infinitum*. However, after the *finite* time that it would have taken the first ball to reach the point 0, had the other balls not been in its way, every ball will have moved briefly, but then have been brought to rest. After all motion has subsided, the energy and momentum of the balls have disappeared without trace.

Figure 1: Collision of an infinite number of identical balls

The conclusion is that the laws of conservation of energy and momentum have been violated. What is more, since classical mechanics is time-reversal invariant, a video recording of the above scenario of collisions, run backwards in time, should also depict a possible mechanical evolution that is consistent with the Newtonian equations. Such an evolution begins with an infinite number of identical balls that are at rest. At a certain moment motion arises spontaneously, out of the origin, and balls move to the right, successively passing on the motion to their rightmost neighbour until the last ball carries off all the energy and momentum. Since the moment at which the motion commences, and its magnitude for that matter, are arbitrary, the system is grossly indeterministic.

What has the Belgian to say to the Basque? Since there are infinitely many balls, all of the same mass, the total mass of all the balls is infinite, in clear violation of condition (C2), so the breakdown of determinism is no surprise. It does not help to remove the restriction that the balls be point masses. For suppose the balls, instead of being point masses, are spheres of geometrically decreasing radii, fitting on a finite line segment (see Figure 2). Then, if the balls are progressively more dense, in such a way that they all have the same mass, the analysis goes through unchanged: an infinite number of elastic collisions leads to the loss of all energy and momentum, and to the breakdown of determinism.

How would it be if the density of the progressively smaller balls were *constant*, so that the masses decrease geometrically? Now each ball is not brought to full rest by collision with its neighbour, so that it retains some kinetic energy after its final collision. Is it possible that the sum of the energies of all the balls, after the infinite sequence of collisions has taken its course, is equal to the initial energy? Indeed that is what happens when the masses decrease in geometrical progression, at any rate if the rules of Newtonian mechanics are followed (Atkinson and Johnson 2009). Energy and momentum are conserved, determinism is inviolate, and sanity seems to have been restored.

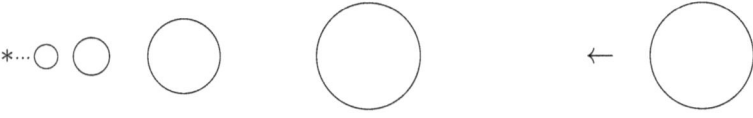

Figure 2: Collision of an infinite number of progressively smaller balls

It turns out that the velocities of the balls are not bounded from above: some of the very tiny balls, indeed all but a finite number of them, acquire speeds in excess of that of light. Since the total mass of all the balls is now finite, condition (C2) is respected by the new system of balls, but how about (C4)? The velocities have now no upper bound, but none of them is actually infinite. Van Bendegem writes that his prohibition (C1)–(C4)

> ... does not imply that masses, forces, accelerations and velocities have finite upper bounds. It may very well be that arbitrarily large values are allowed, only excluding the occurrence of the infinite value itself. (Van Bendegem 1992, p. 33)

Apparently, then, the unboundedness of the velocities does not count as a violation of (C4).

Superluminal velocities were anathema to Albert Einstein, if not to Isaac Newton. What happens if we reconsider the Zeno ball system, with geometrically decreasing masses, according to special relativistic mechanics? When this is done, it is found that the tiny balls have speeds close to, but always less than that of light, as expected. One might also expect the system to obey the law of conservation of energy-momentum, but it does not!

When the energy-momenta of the balls are added up, after all the collisions have taken place, it is found that some energy and some momentum has been lost. Although energy-momentum is conserved after a finite number of elastic collisions, after an infinite number of them this is no longer the case. Moreover determinism has been thrown out of the window too, for the time-reversed scenario involves once more an undetermined time at which energy-momentum is created *ex nihilo* at the origin (Atkinson and Johnson 2009).

How do Van Bendegem's conditions fare in this relativistic setting? (C4) is safe, for all velocities are less than that of light, although they do approach that finite value asymptotically. (C2) is also inviolate, for the sum of the rest masses is finite. In fact the sum of the kinetic masses is finite too, for it turns out that this sum can never exceed the initial energy of the first ball at the beginning of the exercise. How about (C1) and (C3)? Are there any infinite forces or accelerations? In a sense there are, if we stick to what are usually called impulsive forces; but this is surely a red herring. We can easily replace the impulsive forces of collision by finite-range forces (for example, where the range decreases along the line of Zeno balls in such a way that it is never greater than one-third of the initial distance between successive balls), and if we do that, (C1) and (C3) are clearly maintained.

What does this mean? We have managed to respect (C1)–(C4), and yet determinism has fallen by the wayside! Van Bendegem writes:

> Summarizing, what these few examples clearly show is that (C1)–(C4) really are necessary to avoid the conclusion of indeterminism. However, it does not show that the subset of models that satisfies these four principles will coincide with the deterministic models. In other words, what guarantee do we have that indeed all these models are deterministic? (Van Bendegem 1992, pp. 42–43)

So no watertight claim is being made that (C1)–(C4) are sufficient to guarantee determinism, but we do read:

> Although I have not been able —in fact, I am quite skeptical ...that it could be possible at all— to find a formal proof that (C1)–(C4) are sufficient, I did find additional arguments in its favour. (Van Bendegem 1992, p. 43)

We submit that the search might as well be be stopped: (C1)–(C4) are demonstrably insufficient to guarantee determinism, for our relativistic Zeno ball model respects them all, but is indeterministic. We have seen that for Jean Paul each of the conditions (C1)–(C4) is necessary for determinism, thus two questions suggest themselves:

1. Was it essential to employ Einstein's mechanics to show that (C1)–(C4) are insufficient? Can the same thing be done using Newton's mechanics?

2. What must be added to (C1)–(C4) to achieve necessary *and sufficient* conditions for determinism?

As to the first question, we can be brief. Although determinism is respected in classical mechanics when the masses of the balls decrease geometrically, that is not so when the masses decrease more slowly. For example, if the nth ball has a mass that is proportional to $1/n^2$, it can be shown that: (1) the total mass of all the balls is finite, thus safeguarding (C2); (2) energy is not conserved by the infinite set of Zeno balls; and (3) indeterminism reigns (Atkinson and Johnson 2009). (C4) is not violated, since no ball has infinite velocity, even though there is no upper bound on the velocities. The finitude of forces and accelerations can again be assured by the subterfuge of finite-range, instead of impulsive forces. Thus (C1)–(C4) are insufficient as guarantors of determinism, also in classical (Newtonian) mechanics.

The quest for sufficient conditions to assure determinism is more subtle: we will consider the matter at length in the next section.

3 Colliding with an open set

In this section we consider an enrichment of the Zeno ball model in which Van Bendegem anticipated —by six years— a model that Alper and Bridger used to criticize Pérez Laraudogoitia's paper (Alper and Bridger 1998).

Figure 3: VB ball and Zeno balls

As in Figure 1, an infinite number of identical, stationary Zeno balls are placed at the Zeno points $1, \frac{1}{2}, \frac{1}{4}, \frac{1}{8} \ldots$ on a straight line. Van Bendegem took the balls to be of decreasing size and mass, but, as he points out, the same point can be made with identical point masses, so we will first present the argument in those terms. The novel feature is that a further ball, which we shall call the *Van Bendegem ball*, or VB ball for short, is identical to the others and is situated on the line to the left of the origin, 0. It moves with constant speed towards the origin (see Figure 3).

There is no ball at 0, but the origin is a point of accumulation of the locations of the Zeno balls. If the VB ball were to collide with a Zeno ball, it would come to rest, thereby imparting all its energy to the Zeno ball, which would move off with the speed that the VB ball originally had. However, there is no Zeno ball with which it could collide. For suppose, *per impossibile,* that it did collide with one of the Zeno balls. Then it should first have collided with that Zeno ball's immediate lefthand neighbour, and this would have brought it to rest, making the posited collision impossible. Thus the VB ball can collide with none of the Zeno balls. In the absence of any forces other than those arising from collision, the VB ball must continue in its state of constant motion (Newton's first law), thus arriving in a finite time at the Zeno point 1. But this is also impossible, since an infinite number of Zeno balls should have blocked its way.

Such is the scenario sketched by Van Bendegem, and later by Alper and Bridger. The latter authors conclude that what we have called the VB ball must simply cease to exist when it arrives at the origin, since it can neither be stopped there, nor can it progress further, nor indeed can it be anywhere else. Not only has energy disappeared without trace, as in Pérez Laraudogoitia's related problem, but the ball itself has vanished into thin air!

However, Alper and Bridger's system constitutes a logical contradiction. To be precise the following conditions are inconsistent with one another:[1]

1. Stationary balls of unit mass and zero size —mass points— are situated at the Zeno points.

2. A moving ball of unit mass and zero size travels at constant speed, reaching 0 from the left.

3. When the moving mass point occupies the same position as a stationary mass point, it comes to rest, otherwise it continues in its state of constant motion.

4. The moving ball comes to rest before reaching the point 1.

While it is formally possible in classical logic to deduce anything from a contradiction, of course to say anything significant about a physical

---

[1]This, and other inconsistent systems that are isomorphic to it, have been considered by Peijnenburg and Atkinson (2008) and Peijnenburg and Atkinson (2010).

system (however idealized) one should start from a noncontradictory set of statements. Moreover, Alper and Bridger's statements are doubly suspect, for they ask us to believe that there is no trouble until the VB ball reaches the origin, but that the system becomes paradoxical at the moment that the ball reaches that point. However, that is a misreading of the situation: the system as described by the four above conditions is inconsistent *tout court*, not simply at one particular time.

What did Jean Paul Van Bendegem say about the paradox, six years before Alper and Bridger sunk their teeth into the problem? After noting, as above, that the VB ball cannot collide with any of the Zeno balls, he concludes that the VB ball must move through the Zeno balls, which, as he wryly admits, is

> ... an odd conclusion to say the least. (Van Bendegem 1992, p. 40)

He claims

> The solution to [this] problem is easily seen ... Instead of keeping the [Zeno balls] separate, join them. Instead of spheres, turn them into rectangles with the same height H for each one, but with diminishing length L_i. (Van Bendegem 1992, pp. 40–41)

As we noted, Van Bendegem is in the first instance talking about balls of finite, decreasing size, rather than point masses. His model is illustrated in Figure 4.

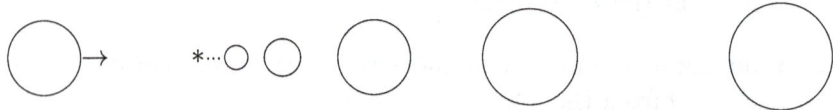

Figure 4: VB ball and an infinite number of progressively smaller balls

Van Bendegem proposes turning the spheres into 'rectangles' —or more properly right parallelepipeds— and then joining them to produce one solid body. He continues:

> Let the sum of all these lengths be L. Then it is obvious what we obtain: a nice rectangle of size L by H ... when we are

talking about an object ... normally speaking, we include the boundary or limit points. (Van Bendegem 1992, p. 41)

It seems to us however that the proposed 'solution' does not address the Van Bendegem-Alper-Bridger paradox at all. The model in which the balls are all joined up is, as a mathematical problem, quite different from the model depicted in Figures 3 and 4. The former model has a straightforward, deterministic solution, whilst the latter is mechanically impossible. It would not help to add a point mass at the origin, the limit point of the Zeno balls. For although the VB ball could then collide with this extra point mass, the latter could not make contact with any of the Zeno balls, and this for the same reason as before.

Finally, Jean Paul the confessed strict finitist adds:

> Note too that these examples illustrate in a quite clear way that not all infinities in classical mechanics are to be eliminated. Limit points are typically products of an infinite process. If one were to insist on the elimination of all infinities, then the proposed solution would be excluded. (Van Bendegem 1992, p. 42)

We are not particularly fond of Alper and Bridger's conclusion that, since the system involves a logically inconsistent set of properties, the ball must spontaneously disappear when it reaches the point of accumulation. A more straightforward way to avoid the inconsistency, in our view, is to ban an actual infinity of balls. If we insist that the number of Zeno balls is only potentially infinite, we can easily answer the question as to what happens to the VB ball. The properties of the (potentially) infinite set of Zeno balls are defined to be the limits (when these exist) of properties of a finite number of balls, as that number increases without bound.

When the Zeno balls are all identical, we find that the VB ball comes to rest. If the masses of the Zeno balls decrease geometrically, the results of numerical calculations show that, when the new ball is light, it rebounds, but if it is sufficiently massive, it either stops, or it merely slows down, i.e. it moves with a reduced positive velocity after all collisions have taken place (Atkinson and Johnson 2010).

In conclusion, we suggest that, in addition to Van Bendegem's four conditions (C1)–(C4) one should add the condition that all actual infinities had better be banned from physics. Indeed, a case could be made that the limitation of mechanics to the potentially infinite provides sufficient means to avoid all the pathological mathematical creatures on Jean Paul Van Bendegem's fascinating roll call.

References

Alper, J. S. and M. Bridger (1998). "Newtonian Supertasks: A Critical Analysis". In: *Synthese* 114, pp. 335–369.

Atkinson, D. and P. W. Johnson (2009). "Nonconservation of Energy and Loss of Determinism. I. Infinitely many colliding balls". In: *Foundations of Physics* 39, pp. 937–957.

— (2010). "Nonconservation of Energy and Loss of Determinism. II. Colliding with an open set". In: *Foundations of Physics* 40, pp. 179–189.

Peijnenburg, J. and D. Atkinson (2008). "Achilles, the Tortoise and Colliding Balls". In: *History of Philosophy Quarterly* 25, pp. 187–201.

— (2010). "Lamps, Cubes, Balls and Walls. Zeno problems and solutions". In: *Philosophical Studies* 150, pp. 49–59.

Pérez Laraudogoitia, J. (1996). "A Beautiful Supertask". In: *Mind* 105, pp. 81–83.

Van Bendegem, J. P. (1992). "How Infinities Cause Problems in Classical Physical Theories". In: *Philosophica* 50, pp. 33–54.

CHAPTER 2

The Consistency of Peano Arithmetic. A Defeasible Perspective

Diderik Batens

Prelude en forme de chicon (Cichorium intybus convar. foliosum)

Jean Paul and me go back for a while. We might be getting older by now. Maybe even grownups, although it is unlikely I'll join him there.

Once a group of students requested me to teach a non-existing logic course—one outside the curriculum. They thought I knew something and wanted me to share it. Jean Paul was among them. He was a character found in groups formed around an original idea. He never took the lead, even when appointed the leader.

Later I supervised his PhD, my second but my first in Ghent. I inherited the supervision from my dearest friend ever, Leo Apostel, who had retired. Actually, Jean Paul restricted the role of his supervisors. He solicited reactions to his ideas, some actually wild. He solicited reactions on drafted chapters, but only at an extremely late stage.

Later on we became friends. I even had some influence on his life, both professional and spiritual. Or so I like to think.

Is it wise to honour him by a paper on arithmetic? After all, he is a mathematician and I am not. Moreover, my paper is not about finitism, which is Jean Paul's main fascination with arithmetic. But why be wise? Especially as the main characters of the subsequent story are close relatives to the finite inconsistent models that meant so much to Jean Paul. The models allowed him to answer "yes yes, sure" to all claims classical mathematicians make on arithmetic, while remaining faithful to his strict finitism. The finite inconsistent models will prove useful for some of my purposes, but are clearly too small for others. I shall slightly extend them for the latter purposes. It is a matter of inserting some numbers between the next to last and the last. The structure remains unchanged.

1 Problem and Aim

What if **PA** (Peano Arithmetic) turns out to be inconsistent? Is there an alternative theory, which is not only non-trivial but also interesting and useful? I shall show there is.

We know from Gödel's second Incompleteness Theorem that if **PA** is consistent, then its consistency can only be proven by means that cannot be represented within **PA**.[1] A slightly different way of stating the matter is by saying that, if **PA** is consistent, then it is impossible to prove its consistency by means that are acceptable for intuitionists.[2]

The two Incompleteness Theorems had consequences for the axiomatic method and for Hilbert's program. Beyond these, the situation led to extremely opposed attitudes. Many keep firmly convinced that **PA** is consistent and regard Gödel's result as a theoretical insight devoid of practical effects. A famous example is Boolos and Jeffrey (1989), in which it is claimed that **PA** is obviously consistent *because it has a model*.[3] This is obviously a poor argument. First, the set of sentences verified by the standard model is not semi-recursive, as Gödel's First Incompleteness Theorem tells us. Every proper superset is inconsistent. The set clearly has no finite models. In general, no known method provides an absolute consistency proof for the truths of the standard model. Moreover, if **PA** is inconsistent, this might have an impact on the method for defining infinite predicative models. Indeed, according

---

[1] Gödel's original paper, Gödel (1931), English translation for example in Gödel (1986), was about a different system. Later the theorem was proved to hold even for weaker systems of arithmetic—see for example Boolos, Burgess, and Jeffrey (2002); Boolos and Jeffrey (1989).

[2] This is the well-known outlook of Kleene (1952). Intuitionist arithmetic is a proper fragment of classical arithmetic. It is sensible, although not faithful to the intuitionists, to say that intuitionist arithmetic is weaker than **PA** because it is closed under a logic that is weaker than Classical Logic.

[3] This claim was removed when Burgess rewrote the book, see for example Boolos, Burgess, and Jeffrey (2002).

to the usual method the standard model is consistent, viz. defines a consistent set of verified sentences, and verifies **PA**.

Others claim that Gödel's Incompleteness Theorems show that **PA** is inconsistent, or rather that it has a recursively enumerable non-trivial extension that is inconsistent as well as natural and correct or 'true'. Among them is Graham Priest, who argued, in Priest (1979) and later for example in Priest (2006), that it is possible to extend *paraconsistent* proof means (by formalizing certain naive proof procedures—basically reasonings with a semantic flavour) in such a way that the Gödel sentence (for **PA**) becomes provable (by a reasoning similar to the one that shows the Gödel sentence true in the standard model). The extended paraconsistent proof means are suggested to define a recursive proof relation and so to assign semi-recursive consequence sets to recursive premise sets. As Gödel's First Incompleteness Theorem is still provable about the system, the resulting arithmetic is inconsistent.

It is not my aim to discuss such claims and attitudes, but I shall nevertheless offer two comments. The first is that, in view of Gödel's theorem, there is no absolute warrant that **PA** is consistent. If it would turn out inconsistent, however, those who consider **PA** as consistent would be left with a useless, viz. trivial, theory.[4] **PA** would not even be a mistaken theory from which one might reason towards a consistent replacement, for reasoning from a trivial theory —actually *the* trivial theory of the considered language— is pointless. The second comment is that the position defended in Priest (1979, 2006) and sketchily summarized in the previous paragraph, does not tell us exactly where the so obtained inconsistent arithmetic leaves us. There are many Gödel sentences even in terms of a single convention on Gödel numbers and there are many possible such conventions, actually infinitely many, and each of them will offer a different variant for every Gödel sentence. So at least from some number on, inconsistencies seem to abound—it seems unavoidable, for example, that the involved Gödel numbers are different from themselves as well as identical to themselves. So the resulting inconsistent arithmetic is apparently not a suitable instrument for practical calculation purposes. Note that this is not a criticism of the claims made in Priest (1979, 2006).

Instead of discussing the considered claims and attitudes, I shall present a result that contributes to our understanding of the problem and has theoretical as well as practical implications. To be more precise,

---

[4] A theory T is a couple $\langle \Gamma, \mathbf{L} \rangle$ in which Γ is a set of formulas (or sentences) and $\mathbf{L}$ is a logic. The set of theorems of $T = \langle \Gamma, \mathbf{L} \rangle$ will be identified with $Cn_{\mathbf{L}}(\Gamma) = \{A \mid \Gamma \vdash_{\mathbf{L}} A\}$, the $\mathbf{L}$-consequence set of Γ, with the language schema (or language) specified where it matters.

I shall present an arithmetic **APA** that has a number of interesting properties.

(1.1) Let **PA** be defined as the deductive closure in terms of **CL** (Classical Logic) of the set Γ_0, which comprises the usual Peano axioms (see Section 4). **APA** will be defined as the deductive closure in terms of an inconsistency-adaptive logic **CLuNs**^{m} (see Section 3) of a set Γ_1, which is **CL**-equivalent to Γ_0.

(1.2) If **PA** is consistent, every theorem of **PA** is a theorem of **APA** and *vice versa*.

(1.3) If **PA** is inconsistent, **PA** is trivial but **APA** is not trivial.

(1.4) Although some **CLuNs**^{m}-consequence sets are Π_1^1-complex, **APA** is well-defined.

(1.5) The computational complexity mentioned in (1.4) concerns the **CLuNs**^{m}-consequence relation. If **PA** is consistent, the set of **APA**-theorems is a semi-recursive set S_0 of **PA**-theorems. If **PA** is inconsistent, S_0 is trivial and the set of **APA**-theorems is a semi-recursive $S_1 \subset S_0$.

(1.6) **PA**-proofs may be algorithmically turned into **APA**-proofs. So if one adopts **APA** instead of **PA**, then, as long as **PA** was not shown inconsistent, one may continue to write **PA**-proofs and to rely upon them. When **PA** is shown to be inconsistent, the algorithm is applied and one goes after the restricted theorems mentioned in (1.8).

(1.7) Every sentence of the language of arithmetic that contains only constant terms—see Section 4—is an **APA**-theorem iff it is a **PA**-theorem. So all arithmetic calculations are safeguarded in **APA** even if **PA** is inconsistent.

(1.8) To every 'desired' **PA**-theorem—see Section 6—seems to correspond a restricted **APA**-theorem, which is a sentence of the language of arithmetic.[5] The full restriction ensures that no term of the theorem refers to the number that is its own successor.[6] Often no restriction is needed or a partial restriction is sufficient.

It does not follow from (1.5) that the set of **APA**-theorems is semi-recursive, because we do not know whether **PA** is consistent or inconsistent. We may set out to collect members of both S_0 and S_1 right now,

---

[5] (1.8) holds on top of (1.2). Intuitively a 'desired' **PA**-theorem is one that is intended to hold (presupposing the consistency of **PA**).

[6] If **PA** is inconsistent, $\exists x\, x' = x$ is an **APA**-theorem, but none of its instances is. Worse, $\neg t' = t$ is provable for every constant term t (and it is provable that $t' = t$ entails triviality). So if **PA** is inconsistent, **APA** is inconsistent and, independently of this, ω-inconsistent.

but as long as we do not find an inconsistency (or an absolute metatheoretic consistency proof), we will not know whether the members of $S_0 - S_1$ are **APA**-theorems.

My aim in the present paper is to show that there is a syntactic theory, **APA**, which has the properties (1.1)–(1.8). So I explicitly leave room for the possibility that Arithmetic (the sentences verified by the standard model) and even **PA** are inconsistent. I am not looking for an inconsistent extension of Arithmetic, but for a theory that is as consistent as possible with respect to **PA**. The properties (1.1)–(1.8) are not exaggerated. Those who work through the paper will see that the results are actually more impressive than they appear at this point. The reader should be warned, however. Please read the next paragraph carefully.

This paper relies on somewhat unusual presuppositions, especially for the reader not acquainted with paraconsistency and for the reader not acquainted with defeasible reasoning. A related but different matter is that metatheoretic proofs often presuppose classical means of proof (classical logic, classical set theory, sometimes up to classical arithmetic), even when the authors claim (and believe) that their metatheoretic proofs do not. So I shall try to write as transparently as possible and I shall explicitly discuss the presuppositions of the results in Section 6.

2 A Very Strong Paraconsistent Logic

A very strong paraconsistent logic, which happens to be extremely popular, is called **CLuNs** in Ghent.[7] This logic will be formulated in the standard language schema $\mathcal{L}_s$; let $\mathcal{F}_s$ be its set of formulas and $\mathcal{W}_s$ its set of closed formulas. A formula is *atomic* iff it contains no logical symbols except possibly for identity. Let $\mathcal{F}_s^a$ and $\mathcal{W}_s^a$ be the atomic formulas in $\mathcal{F}_s$ and $\mathcal{W}_s$ respectively.

In the sequel, all axioms will be closed formulas or closed sentences. Similarly all formulas mentioned in the clauses of semantic systems will be closed. Where A is an open formula, let $\exists A$ [$\forall A$] denote the *existential closure* [*universal closure*] of A, viz. A preceded by an existential quantifier [universal quantifier] over every variable that occurs free in A.[8]

---

[7]**CLuNs** is known under a multiplicity of names. To the best of my knowledge, the propositional version appears first in Schütte (1960). Further useful references to studies of **CLuNs** and of its fragments are Arieli, Avron, and Zamansky (2011); Avron (1986, 1991); Batens (1980); Batens and De Clercq (2004); Carnielli, Marcos, and Amo (2001); D'Ottaviano (1982, 1985a,b, 1987); D'Ottaviano and Epstein (1988); Esser (2003); Priest (1979, 2006); Smirnova (2000). Proofs for results mentioned in this section can be found in Batens (201x); Batens and De Clercq (2004).

[8]Some examples: $\exists((p \wedge Pa) \wedge \exists x Q x)$ is $(p \wedge Pa) \wedge \exists x Q x$ and $\exists((Ra \wedge Pz) \wedge$

Let **CL$^+$** be an axiom system for full positive **CL** in which all logical symbols are implicitly defined. An axiom system for (a particular version of) **CLuNs** is obtained by extending **CL$^+$** with excluded middle, $A \vee \neg A$,[9] with full RoI (Replacement of Identicals), $\alpha = \beta \supset (A(\alpha) \supset A(\beta))$,[10] and with the following negation reducing axiom schemas: $\neg\neg A \equiv A$, $\neg(A \supset B) \equiv (A \wedge \neg B)$, $\neg(A \wedge B) \equiv (\neg A \vee \neg B)$, $\neg(A \vee B) \equiv (\neg A \wedge \neg B)$, $\neg(A \equiv B) \equiv ((A \vee B) \wedge (\neg A \vee \neg B))$, $\neg \forall \alpha A \equiv \exists \alpha \neg A$, and $\neg \exists \alpha A \equiv \forall \alpha \neg A$. Note that the equivalence is detachable but not contraposable.

Here are a few typical properties of **CLuNs**:

(2.1) $A \wedge \neg A \nvdash_{\mathbf{CLuNs}} B$

(2.2) $A \supset B, A \vdash_{\mathbf{CLuNs}} B$

(2.3) $A \supset B, \neg B \nvdash_{\mathbf{CLuNs}} \neg A$, but $A \supset B, \neg B \vdash_{\mathbf{CLuNs}} \neg A \vee (B \wedge \neg B)$

(2.4) $A \vee B, \neg A \nvdash_{\mathbf{CLuNs}} B$, but $A \vee B, \neg A \vdash_{\mathbf{CLuNs}} B \vee (A \wedge \neg A)$

(2.5) $\vdash_{\mathbf{CLuNs}} \neg(p \supset q) \equiv (p \wedge \neg q)$, but $\nvdash_{\mathbf{CLuNs}} (p \supset q) \equiv \neg(p \wedge \neg q)$. Indeed, $\vdash_{\mathbf{CLuNs}} \neg(p \wedge \neg q) \equiv (\neg p \vee q)$ whereas $\nvdash_{\mathbf{CLuNs}} (p \supset q) \equiv (\neg p \vee q)$.

Different styles of semantic systems for **CLuNs** are described in Batens and De Clercq (2004). Here I shall present yet another deterministic semantics. I first introduce pseudo-languages in order to simplify the description of the semantics. Consider a set of sets $\mathcal{O}$ of pseudo-constants. The sets $\mathcal{O}$ are distinguished from each other by, say, superscripts, which I shall write invisibly in the sequel. For every **CLuNs**-model M some set $\mathcal{O}$ has at least the cardinality of M. For each $\mathcal{O}$, a pseudo-language schema $\mathcal{L}_\mathcal{O}$ is defined exactly like $\mathcal{L}_s$, except that the denumerable set $\mathcal{C}$ of constants is replaced by the set $\mathcal{C} \cup \mathcal{O}$. Let $\mathcal{W}_\mathcal{O}$ be the set of closed pseudo-formulas of $\mathcal{L}_\mathcal{O}$ and let $\mathcal{W}_\mathcal{O}^a$ be the atomic members of $\mathcal{L}_\mathcal{O}$.

A **CLuNs**-model $M = \langle D, v \rangle$, in which D is a non-empty set and v an assignment function, is an interpretation of $\mathcal{W}_\mathcal{O}$ and hence of $\mathcal{W}_s$. The assignment function v is defined by:[11]

C1 $v \colon \mathcal{W}_\mathcal{O} \to \{0, 1\}$

C2 $v \colon \mathcal{C} \cup \mathcal{O} \to D$ (where $D = \{v(\alpha) \mid \alpha \in \mathcal{C} \cup \mathcal{O}\}$)

C3 $v \colon \mathcal{P}^r \to \wp(D^r)$

$(\exists x Q x \wedge R y))$ is $\exists y \exists z ((R a \wedge P z) \wedge (\exists x Q x \wedge R y))$.

[9] The formulas $A \vee \neg A$ and $(\neg A \supset A) \supset A$ are **CL$^+$**-equivalent.

[10] This is not the place to discuss whether RoI is a theorem of **CL$^+$**.

[11] By the restriction in C2, the couple $\langle D, v \rangle$ is not a **CLuNs**-model if $D \neq \{v(\alpha) \mid \alpha \in \mathcal{C} \cup \mathcal{O}\}$. In C3, $\wp(D^r)$ is the power set of the r-th Cartesian product of D.

For every $A \in \mathcal{W}_\mathcal{O}^a$, $[\![A]\!]$ is the smallest set S fulfilling: (i) $A \in S$, and (ii) if $B(\alpha) \in S$ and $v(\beta) = v(\alpha)$, then $B(\beta) \in S$.[12] The valuation function $v_M : \mathcal{W}_\mathcal{O} \to \{0, 1\}$ determined by M is defined as follows:

$\mathrm{C}\mathcal{S}$ where $A \in \mathcal{S}$, $v_M(A) = v(A)$
$\mathrm{C}\mathcal{P}^r$ $v_M(\pi^r \alpha_1 \ldots \alpha_r) = 1$ iff $\langle v(\alpha_1), \ldots, v(\alpha_r) \rangle \in v(\pi^r)$
$\mathrm{C}=$ $v_M(\alpha = \beta) = 1$ iff $v(\alpha) = v(\beta)$
$\mathrm{C}\supset$ $v_M(A \supset B) = 1$ iff $v_M(A) = 0$ or $v_M(B) = 1$
$\mathrm{C}\wedge$ $v_M(A \wedge B) = 1$ iff $v_M(A) = 1$ and $v_M(B) = 1$
$\mathrm{C}\vee$ $v_M(A \vee B) = 1$ iff $v_M(A) = 1$ or $v_M(B) = 1$
$\mathrm{C}\equiv$ $v_M(A \equiv B) = 1$ iff $v_M(A) = v_M(B)$
$\mathrm{C}\forall$ $v_M(\forall \alpha A(\alpha)) = 1$ iff $\{v_M(A(\beta)) \mid \beta \in \mathcal{C} \cup \mathcal{O}\} = \{1\}$
$\mathrm{C}\exists$ $v_M(\exists \alpha A(\alpha)) = 1$ iff $1 \in \{v_M(A(\beta)) \mid \beta \in \mathcal{C} \cup \mathcal{O}\}$
$\mathrm{C}\neg$ Where $A \in \mathcal{W}_\mathcal{O}^a$, $v_M(\neg A) = 1$ iff $v_M(A) = 0$ or $v(\neg B) = 1$ for a $B \in [\![A]\!]$.
$\mathrm{C}\neg\neg$ $v_M(\neg\neg A) = v_M(A)$
$\mathrm{C}\neg\supset$ $v_M(\neg(A \supset B)) = v_M(A \wedge \neg B)$
$\mathrm{C}\neg\wedge$ $v_M(\neg(A \wedge B)) = v_M(\neg A \vee \neg B)$
$\mathrm{C}\neg\vee$ $v_M(\neg(A \vee B)) = v_M(\neg A \wedge \neg B)$
$\mathrm{C}\neg\equiv$ $v_M(\neg(A \equiv B)) = v_M((A \vee B) \wedge (\neg A \vee \neg B))$
$\mathrm{C}\neg\forall$ $v_M(\neg \forall \alpha A(\alpha)) = v_M(\exists \alpha \neg A(\alpha))$
$\mathrm{C}\neg\exists$ $v_M(\neg \exists \alpha A(\alpha)) = v_M(\forall \alpha \neg A(\alpha))$

For all $\Gamma \subseteq \mathcal{W}_s$ and $A \in \mathcal{W}_s$, $M \Vdash A$ (M verifies A) iff $v_M(A) = 1$; $M \Vdash \Gamma$ (M verifies Γ) iff $M \Vdash B$ for all $B \in \Gamma$; $\Gamma \vDash_{\mathbf{CLuNs}} A$ (A is a **CLuNs**-semantic consequence of Γ) iff $M \Vdash A$ whenever $M \Vdash \Gamma$; $\vDash_{\mathbf{CLuNs}} A$ (A is **CLuNs**-valid) iff $M \Vdash A$ for all **CLuNs**-models M. Note that the set of consistent **CLuNs**-models forms a semantics for **CL**.

It is provable that $\Gamma \vdash_{\mathbf{CLuNs}} A$ iff $\Gamma \vDash_{\mathbf{CLuNs}} A$. In view of this, I shall freely switch between the syntax and semantics, for example in proving a claim about derivability in terms of models. I shall proceed similarly for the adaptive logic **CLuNs**^{m} and for the arithmetic **APA**.

Here are some further (syntactic) properties of **CLuNs**:

(2.6) $\vdash_{\mathbf{CLuNs}} \neg \alpha = \beta \equiv \neg \beta = \alpha$.[13]
(2.7)   There are **CL**-models of $\Gamma \subseteq \mathcal{W}_s$ iff Γ is consistent whereas there are **CLuNs**-models of all $\Gamma \subseteq \mathcal{W}_s$.

---

[12] In (ii), $\alpha, \beta \in \mathcal{C} \cup \mathcal{O}$ and $B \in \mathcal{W}_\mathcal{O}^a$. If A is a sentential letter, then $[\![A]\!] = \{A\}$.

[13] From $\neg \alpha = \beta, \beta = \alpha \vdash \neg \beta = \alpha$ (by RoI) and the theorems $\neg \neg \beta = \alpha \equiv \beta = \alpha$ and $(\neg \neg \beta = \alpha \supset \neg \beta = \alpha) \supset \neg \beta = \alpha$. Semantically: if $v_M(\neg \alpha = \beta) = 1$ and $v(\alpha) = v(\beta)$, then $\alpha = \beta \in [\![\beta = \alpha]\!]$, whence $v_M(\neg \beta = \alpha) = 1$; if $v_M(\neg \alpha = \beta) = 1$ and $v(\alpha) \neq v(\beta)$, then $v_M(\neg \beta = \alpha) = 1$. **CLuNs** has many other unexpected properties, which we fortunately don't need in this paper.

(2.8) $\exists(A_1 \wedge \neg A_1) \vee \ldots \vee \exists(A_n \wedge \neg A_n) \in Cn_{\mathbf{CL}}(\Gamma)$ iff there are $B_1, \ldots, B_m$ ($m > 0$) such that $\exists(B_1 \wedge \neg B_1) \vee \ldots \vee \exists(B_m \wedge \neg B_m) \in Cn_{\mathbf{CLuNs}}(\Gamma)$ (Γ is **CL**-inconsistent iff it is **CLuNs**-inconsistent).

(2.9) $\exists(A_1 \wedge \neg A_1) \vee \ldots \vee \exists(A_n \wedge \neg A_n) \in Cn_{\mathbf{CLuNs}}(\Gamma)$ iff there are $B_1, \ldots, B_m \in \mathcal{F}_s^a$ ($m > 0$) such that $\exists(B_1 \wedge \neg B_1) \vee \ldots \vee \exists(B_m \wedge \neg B_m) \in Cn_{\mathbf{CLuNs}}(\Gamma)$.

(2.10) Some contradictions[14] are not **CLuNs**-derivable from each other. For example $p \wedge \neg p \nvdash_{\mathbf{CLuNs}} q \wedge \neg q$ and $q \wedge \neg q \nvdash_{\mathbf{CLuNs}} p \wedge \neg p$. However, some contradictions are (contextually or absolutely) derivable from each other. Examples are $p \wedge \neg p \vdash_{\mathbf{CLuNs}} (p \wedge p) \wedge \neg (p \wedge p)$ and $p \wedge \neg p, q \vdash_{\mathbf{CLuNs}} (p \wedge q) \wedge \neg(p \wedge q)$.

(2.11) With respect to any $A \in \mathcal{W}_s$, some **CLuNs**-models verify A and falsify $\neg A$ (A consistently true), some falsify $\neg A$ and verify A (A consistently false), and some verify both A and $\neg A$ (A inconsistent). These three subsets form a partition of the **CLuNs**-models *with respect to A*. Different formulas obviously lead to different partitions and for some formulas one of the sets is empty.

(2.12) Where $\Gamma \cup \{A\} \subseteq \mathcal{W}_s$, $\Gamma \vdash_{\mathbf{CL}} A$ iff there are $B_1, \ldots, B_n \in \mathcal{F}_s^a$ ($n \geq 0$), which are subformulas of members of $\Gamma \cup \{A\}$ and are such that $\Gamma \vdash_{\mathbf{CLuNs}} A \vee \exists(B_1 \wedge \neg B_1) \vee \ldots \vee \exists(B_n \wedge \neg B_n)$.

(2.12) states the so-called Derivability Adjustment Theorem, which is important for the adaptive logic that extends **CLuNs**—see Section 3.

In the language of arithmetic $\mathcal{L}_a$—see Section 4—occur no predicates but occur two binary functions and one unary function. As everyone knows, a language containing n-ary functions can be rephrased in terms of a language containing $(n+1)$-ary predicates for which a specific axiom or premise is introduced. Thus if $f(a, b) = c$ is replaced by $Pabc$ then we need $\forall z \forall y \forall x \forall w((Pwxy \wedge Pwxz) \supset y = z)$. I leave to the reader the obvious task of spelling out **CLuNs** defined over $\mathcal{L}_a$.

Excursion In many papers on adaptive logics, the language is extended with classical logical symbols. These are separate symbols, like $\check{\neg}$ and $\check{=}$, that have the same meaning as the standard symbols have in **CL**. The classical symbols are used for technical reasons only. In general, they provide insights and simplify certain things, especially

---

[14] Strictly speaking, I mean existentially closed contradictions: $\exists(A \wedge \neg A)$. The matter is not very important for formulas in general, because $\exists(A \wedge \neg A) \vdash_{\mathbf{CLuNs}} \exists(A \wedge \neg A) \wedge \neg\exists(A \wedge \neg A)$ in view of $\vdash_{\mathbf{CLuNs}} \neg\exists(A \wedge \neg A)$, which comes to $\vdash_{\mathbf{CLuNs}} \forall(A \vee \neg A)$. If, however, one considers only contradictions in which A is atomic, one needs to include existentially closed contradictions.

metatheoretic proofs. I shall not introduce them here for two reasons. First, although the classical symbols (in the present paper especially classical negation) provide certain insights, this gain has a price: the reader should master their specific character and use. Next, employing classical negation is unacceptable for dialetheists and I want to avoid the false suggestion that my approach requires classical negation.

3 An Inconsistency-Adaptive Logic

Adaptive logics are formally decent characterizations of defeasible reasoning forms. Inconsistency-adaptive logics more specifically interpret inconsistent theories as consistently as possible. Consider a simple propositional premise set,

$$\Gamma^\dagger = \{p \vee q, \neg p, \neg q, p \vee r, \neg s, s \vee t\}.$$

The **CL** consequence set of $\Gamma^\dagger$ is trivial: $Cn_{\mathbf{CL}}(\Gamma^\dagger) = \mathcal{W}_s$. The **CLuNs** consequence set of $\Gamma^\dagger$ is not trivial; for example $r, t \notin Cn_{\mathbf{CLuNs}}(\Gamma^\dagger)$. This is easily seen in terms of the semantics: for some **CLuNs**-models M of $\Gamma^\dagger$, $v_M(p) = v_M(\neg p) = v_M(s) = v_M(\neg s) = 1$ and $v_M(r) = v_M(t) = 0$.

In this section I present an inconsistency-adaptive logic $\mathbf{CLuNs}^m$. It holds that $Cn_{\mathbf{CLuNs}}(\Gamma^\dagger) \subset Cn_{\mathbf{CLuNs}^m}(\Gamma^\dagger)$. This is easily understood in terms of the **CLuNs**-semantics. All **CLuNs**-models of $\Gamma^\dagger$ verify either $p \wedge \neg p$ or $q \wedge \neg q$ and most of them verify many other inconsistencies. While $Cn_{\mathbf{CLuNs}}(\Gamma^\dagger)$ comprises the formulas that are verified by all **CLuNs**-models of $\Gamma^\dagger$, $Cn_{\mathbf{CLuNs}^m}(\Gamma^\dagger)$ comprises the formulas verified by the minimal abnormal **CLuNs**-models of $\Gamma^\dagger$—these are the models M for which there is no **CLuNs**-model M' of $\Gamma^\dagger$ such that the set of inconsistencies verified by M' is a proper subset of the set of inconsistencies verified by M.[15] Please check that these minimal abnormal models verify either $p \wedge \neg p$ or $q \wedge \neg q$ but not both and verify no other inconsistency. So all minimal abnormal **CLuNs**-models of $\Gamma^\dagger$ verify t because none of them verifies $s \wedge \neg s$. So $t \in Cn_{\mathbf{CLuNs}^m}(\Gamma^\dagger) - Cn_{\mathbf{CLuNs}}(\Gamma^\dagger)$.[16]

Let us now move to the official formulation of $\mathbf{CLuNs}^m$. An adaptive logic in standard format is a triple consisting of (i) a *lower limit logic* **LLL**: this is a reflexive, transitive, monotonic, uniform, and compact logic, for which there is a positive test (which means that it assigns semi-recursive consequence sets to recursive premise sets), (ii) a set of abnormalities Ω, characterized by a (possibly restricted) logical form,

---

[15] The statement is accurate because **CLuNs** is sound and complete with respect to its semantics and because the same holds for $\mathbf{CLuNs}^m$.

[16] Note that $r \notin Cn_{\mathbf{CLuNs}^m}(\Gamma^\dagger)$ because some minimal abnormal **CLuNs**-models of $\Gamma^\dagger$ verify $p \wedge \neg p$ and falsify r.

(iii) an adaptive *strategy* (Reliability or Minimal Abnormality)—the effect of this becomes clear below. The upper limit logic **ULL** of an adaptive logic is obtained by extending **LLL** with an axiom or rule in such a way that every member of Ω entails triviality.

It is instructive to compare, on the one hand, the relation between **LLL** and **ULL** and, on the other hand, the relation between **LLL** and the adaptive logic **AL**. **ULL** extends **LLL** by validating some rules not validated by **LLL**. **AL**, however, extends **LLL** by validating certain *applications* of those **ULL**-rules. Which applications are validated depends on *the contents of the premises*—in the case of $Cn_{\mathbf{CLuNs}^m}(\Gamma^\dagger)$, the fact that $(p \wedge \neg p) \vee (q \wedge \neg q)$ is the only disjunction of abnormalities that is **CLuNs**-derivable from $\Gamma^\dagger$.

In case of the inconsistency-adaptive **CLuNs**^{m} (i) the lower limit logic is **CLuNs**, (ii) the set of abnormalities: $\Omega = \{\exists(A \wedge \neg A) \mid A \in \mathcal{F}_s^a\}$,[17] and (iii) the Minimal Abnormality strategy. The upper limit logic of **CLuNs**^{m} is **CL**.

If an adaptive logic is in standard format, as **CLuNs**^{m} is, the standard format itself (and no specific properties of the logic) provides it with: (i) a dynamic proof theory,[18] (ii) a semantics (a selected set of **LLL**-models), and (iii) nearly all the metatheory, including soundness, completeness, Tarski-like properties, and many other properties.

The dynamic proof theory is determined by a set of rules on the one hand and a marking definition on the other hand. Lines of a(n annotated) dynamic proof have four elements: a line number, a formula, a justification, and a *condition*, which is a finite subset of Ω. Let Γ be the premise set, let $Dab(\Delta)$ abbreviate the disjunction of the members of the finite $\Delta \subset \Omega$, and let

$$A \quad \Delta$$

abbreviate that A occurs in the proof on the condition Δ. The generic rules of **CLuNs**^{m} are:

---

[17]These abnormalities are logical falsehoods of **CL**, but this is not the case for all adaptive logics. In general "abnormality" should be considered as a technical term.

[18]The dynamic proof theory led to the 'discovery' of adaptive logics and to most interesting new results.

Prem	If $A \in \Gamma$:	...	...
		A	$\emptyset$

RU	If $A_1, \ldots, A_n \vdash_{\mathbf{CLuNs}} B$:	A_1	Δ_1
		...	...
		A_n	Δ_n
		B	$\Delta_1 \cup \ldots \cup \Delta_n$

RC	If $A_1, \ldots, A_n \vdash_{\mathbf{CLuNs}} B \vee Dab(\Theta)$:	A_1	Δ_1
		...	...
		A_n	Δ_n
		B	$\Delta_1 \cup \ldots \cup \Delta_n \cup \Theta$

Note that the conditional rule RC is the only one that introduces formulas into a condition. The rules refer to the lower limit logic and to the set of abnormalities. They are independent of the strategy. It can be shown that a line at which A is derived on the condition Δ occurs in a **CLuNs**^{m}-proof from Γ iff the formula $A \vee Dab(\Delta)$ occurs in a **CLuNs**-proof from Γ—note that the latter proof lines have no condition.

A proof is seen as a chain of *stages* and a stage is a sequence of lines; the zeroth stage is the empty sequence. Whenever a line is added in accordance to the rules, the resulting sequence of lines forms the next stage of the proof.[19] Where $Dab(\Delta_1), \ldots, Dab(\Delta_n)$ are the disjunctions of abnormalities that occur on the empty condition in stage s of the considered **CLuNs**^{m}-proof from Γ, $\Phi_s(\Gamma)$ is the set of minimal choice sets of $\Sigma = \{\Delta_1, \ldots, \Delta_n\}$.[20] The *marking definition* of **CLuNs**^{m} may be phrased as follows: where A is derived on the condition Δ at line i, line i is *unmarked* at stage s iff (i) there is a $\varphi \in \Phi_s(\Gamma)$ for which $\Delta \subseteq \Omega - \varphi$ and (ii) for every $\varphi \in \Phi_s(\Gamma)$, there is a line at which A is derived on a condition Θ for which $\Theta \subseteq \Omega - \varphi$. I shall comment on this definition after introducing the semantics.

Formulas of lines that are unmarked at stage s are considered as derived from the premises in view of the insights provided by the stage, viz. in view of the disjunctions of abnormalities that are derived on the empty condition at stage s. Formulas of lines that are marked at stage s are considered as not derived from the premises in view of the insights

---

[19] The rules do not forbid to *insert* a line l as long as the justification of a line l refers only to lines that precede l in the resulting sequence.

[20] A choice set of $\Sigma = \{\Delta_1, \ldots, \Delta_n\}$ is a set that contains an element of each member of Σ. A choice set φ of Σ is *minimal* iff no choice set φ' of Σ is such that $\varphi' \subset \varphi$. Replacing "disjunctions of abnormalities" by "minimal disjunctions of abnormalities" has no effect on $\Phi_s(\Gamma)$.

provided by the stage. The connected notion is derivability at a stage. As a formula may be derivable at a stage, non-derivable at a later stage, and again derivable at a still later stage, we also want a stable notion of derivability, which is called *final derivability*. Let $\Phi(\Gamma)$ be defined like $\Phi_s(\Gamma)$ but with reference to the disjunctions of abnormalities, $Dab(\Delta_1)$, $Dab(\Delta_2)$, ..., that are **CLuNs**-derivable from Γ. As infinitely many minimal disjunctions of abnormalities may be **CLuNs**-derivable from Γ, it is better to present a definition that does not refer to a single stage or even a single proof. A formally correct definition reads as follows:

Definition 1 $\Gamma \vdash_{\mathbf{CLuNs}^m} A$ *(A is finally $\mathbf{CLuNs}^m$-derivable from Γ) iff (i) there is a stage s of a $\mathbf{CLuNs}^m$-proof from Γ in which A occurs as the formula of an unmarked line l and (ii) whenever line l is marked in a stage s' that extends stage s, then stage s' may be extended in such a way that line l is unmarked.*

It is instructive to consider a simple propositional example proof—$\mathbf{CLuNs}^m$ is decidable at the propositional level. Let $\Gamma^{\ddagger} = \{\neg p \wedge \neg q, p \vee q, p \vee r, q \vee s\}$ be the premise set.

1	$\neg p \wedge \neg q$	Premise	$\emptyset$	
2	$p \vee q$	Premise	$\emptyset$	
3	$p \vee r$	Premise	$\emptyset$	
4	$q \vee s$	Premise	$\emptyset$	
5	r	1, 3; RC	$\{p \wedge \neg p\}$	$\checkmark^7$
6	s	1, 4; RC	$\{q \wedge \neg q\}$	$\checkmark^7$
7	$(p \wedge \neg p) \vee (q \wedge \neg q)$	1, 2; RU	$\emptyset$	
8	$r \vee s$	5; RU	$\{p \wedge \neg p\}$	$\checkmark^7$

At this stage, every line with a non-empty condition is marked. So the derivable formulas are those derivable by **CLuNs**. In other words, going adaptive did, in view of the insights provided by this stage of the proof, not lead to any further consequence. This is changed by adding one line—I rewrite the proof from line 5 on.

5	r	1, 3; RC	$\{p \wedge \neg p\}$	$\checkmark^7$
6	s	1, 4; RC	$\{q \wedge \neg q\}$	$\checkmark^7$
7	$(p \wedge \neg p) \vee (q \wedge \neg q)$	1, 2; RU	$\emptyset$	
8	$r \vee s$	5; RU	$\{p \wedge \neg p\}$	
9	$r \vee s$	6; RU	$\{q \wedge \neg q\}$	

Note that, at the present stage (viz. stage 9), the added line 9 is unmarked, and line 8 is also unmarked. Why is this? Well, $\Phi_9(\Gamma^{\ddagger}) = \{\{p \wedge \neg p\}, \{q \wedge \neg q\}\}$, and the same formula, $r \vee s$, occurs at a line with condition $\{p \wedge \neg p\}$, which does not overlap with one member of $\Phi_9(\Gamma^\ddagger)$, and also occurs at a line with condition $\{q \wedge \neg q\}$, which does not overlap with the other member of $\Phi_9(\Gamma^\ddagger)$. Incidentally, it is easily seen that $\Phi(\Gamma^\ddagger) = \Phi_9(\Gamma^\ddagger) = \Phi_7(\Gamma^\ddagger) = \{\{p \wedge \neg p\}, \{q \wedge \neg q\}\}$. So lines 8 and 9 will be unmarked in every extension of stage 9 and $r \vee s$ is finally **CLuNs**^{m}-derivable from $\Gamma^\ddagger$.

The proof theory of **CLuNs**^{m} is somewhat complicated because of the marking definition. The **CLuNs**^{m}-semantics, however, is extremely simple. Let $Ab(M)$ be the set of abnormalities verified by the **CLuNs**-model M. A **CLuNs**-model M of Γ is a *minimally abnormal model* of Γ iff there is no **CLuNs**-model M' of Γ such that $Ab(M') \subset Ab(M)$.

Definition 2 $\Gamma \vDash_{\mathbf{CLuNs}^m} A$ *(A is a **CLuNs**^{m}-semantic consequence of Γ) iff every minimally abnormal model of Γ verifies A.*

It is easily seen (and not difficult to prove) that M is a minimally abnormal model of Γ iff $Ab(M) \in \Phi(\Gamma)$. This clarifies the marking definition. If the minimal disjunctions of abnormalities that occur at unmarked lines of stage s of a proof from Γ are the only minimal disjunctions of abnormalities that are **CLuNs**-derivable from Γ, then the formulas of unmarked lines of stage s are verified by all minimal abnormal models of Γ whereas the formulas of marked lines of stage s are falsified by some minimal abnormal models of Γ.

It is also useful to apply this to the example proof from $\Gamma^\ddagger$. The minimal abnormal models of $\Gamma^\ddagger$ that falsify $p \wedge \neg p$ verify r and the minimal abnormal models of $\Gamma^\ddagger$ that falsify $q \wedge \neg q$ verify s. So all minimal abnormal models of $\Gamma^\ddagger$ verify $r \vee s$.

I now list some properties of **CLuNs**^{m} that we shall need in the sequel. Proofs of all of them and of many other properties occur in Batens (2007, 201x).

(3.1) $\Gamma \vdash_{\mathbf{CLuNs}^m} A$ iff $\Gamma \vDash_{\mathbf{CLuNs}^m} A$.
(3.2) $Cn_{\mathbf{CLuNs}}(\Gamma) \subseteq Cn_{\mathbf{CLuNs}^m}(\Gamma) \subseteq Cn_{\mathbf{CL}}(\Gamma)$
(3.3) If Γ is consistent, then $Cn_{\mathbf{CLuNs}^m}(\Gamma) = Cn_{\mathbf{CL}}(\Gamma)$.
(3.4) If Γ is inconsistent, viz. $Cn_{\mathbf{CL}}(\Gamma) = \mathcal{W}_s$, and $Cn_{\mathbf{CLuNs}}(\Gamma) \neq \mathcal{W}_s$, then $Cn_{\mathbf{CLuNs}^m}(\Gamma) \subset Cn_{\mathbf{CL}}(\Gamma)$.
(3.5) If there is a finite $\Delta \subset \Omega$ such that, for all $\varphi \in \Phi(\Gamma)$, $\Delta - \varphi \neq \emptyset$, then $Cn_{\mathbf{CLuNs}}(\Gamma) \subset Cn_{\mathbf{CLuNs}^m}(\Gamma)$.
(3.6) $Dab(\Delta) \in Cn_{\mathbf{CLuNs}}(\Gamma)$ iff $Dab(\Delta) \in Cn_{\mathbf{CLuNs}^m}(\Gamma)$.
(3.7) If $Cn_{\mathbf{CLuNs}}(\Gamma)$ is non-trivial, then $Cn_{\mathbf{CLuNs}^m}(\Gamma)$ is non-trivial (**Reassurance**).
(3.8) $Cn_{\mathbf{CLuNs}}(Cn_{\mathbf{CLuNs}^m}(\Gamma)) = Cn_{\mathbf{CLuNs}^m}(\Gamma)$ (**LLL-closure**).

(3.9) If $\Gamma' \subseteq Cn_{\mathbf{CLuNs}^m}(\Gamma)$, then $Dab(\Delta) \in Cn_{\mathbf{CLuNs}}(\Gamma \cup \Gamma')$ iff $Dab(\Delta) \in Cn_{\mathbf{CLuNs}}(\Gamma)$.

(3.10) $\Gamma \subseteq Cn_{\mathbf{CLuNs}^m}(\Gamma)$ (Reflexivity).

(3.11) If $\Gamma' \subseteq Cn_{\mathbf{CLuNs}^m}(\Gamma)$, then $Cn_{\mathbf{CLuNs}^m}(\Gamma \cup \Gamma') \subseteq Cn_{\mathbf{CLuNs}^m}(\Gamma)$ (Cumulative Transitivity).

(3.12) If $\Gamma' \subseteq Cn_{\mathbf{CLuNs}^m}(\Gamma)$, then $Cn_{\mathbf{CLuNs}^m}(\Gamma) \subseteq Cn_{\mathbf{CLuNs}^m}(\Gamma \cup \Gamma')$ (Cumulative Monotonicity).

(3.13) If $\Gamma' \subseteq Cn_{\mathbf{CLuNs}^m}(\Gamma)$, then $Cn_{\mathbf{CLuNs}^m}(\Gamma \cup \Gamma') = Cn_{\mathbf{CLuNs}^m}(\Gamma)$ (Cumulative Indifference).

(3.14) $Cn_{\mathbf{CLuNs}^m}(Cn_{\mathbf{CLuNs}^m}(\Gamma)) = Cn_{\mathbf{CLuNs}^m}(\Gamma)$ (Fixed Point).

(3.15) For some Γ, the set $Cn_{\mathbf{CLuNs}^m}(\Gamma)$ is not recursively enumerable. However, it is obviously well-defined.

4 Adaptive Peano Arithmetic

There are a few changes from this section on. The prime, as in x', is the successor function and the letter t functions as a variable for terms. The unqualified "model" will always refer to a **CLuNs**-model. Once Γ_1 is introduced, "**APA**-model" will refer to a **CLuNs**^{m}-model of Γ_1.

Let $\mathcal{L}_a$ be the language of arithmetic. This language's symbols are the constant 0, the set $\mathcal{V}$ of variables of $\mathcal{L}_s$, the unary successor function ', the binary functions $+$ and $\times$, and the logical symbols of $\mathcal{L}_s$, including identity. The set of closed sentences of $\mathcal{L}_a$ will be $\mathcal{W}_a$; $\mathcal{W}_a^a$ will comprise the atomic members of $\mathcal{W}_a$. Let $0^{(n)}$ denote the string consisting of 0 followed by n symbols '. Also, I shall sometimes write decimal numbers instead of numbers in official notation; for example 7 instead of $0'''''''$.

From this section on, $\mathcal{L}_\mathcal{O}$ will refer to a (suitable) pseudo-language that extends $\mathcal{L}_a$ with a set $\mathcal{O}$ of pseudo-constants, which will be used exclusively in a semantic context; analogously for $\mathcal{W}_\mathcal{O}$, $\mathcal{W}_\mathcal{O}^a$, and similar expressions. Let $\mathcal{T}$ be the set of *terms*, defined as the smallest set S fulfilling (i) $\mathcal{V} \cup \{0\} \cup \mathcal{O} \subseteq S$ and (ii) if $t_1, t_2 \in S$, then $t_1', (t_1 + t_2), (t_1 \times t_2) \in S$. We shall need four sets of terms, obtained by varying on (i) in the previous definition:

(i)	$\mathcal{V} \cup \{0\} \subseteq S$	$\mathcal{V} \cup \{0\} \cup \mathcal{O} \subseteq S$	$\{0\} \subseteq S$	$\{0\} \cup \mathcal{O} \subseteq S$
set	$\mathcal{T}_a$	$\mathcal{T}_\mathcal{O}$	$\mathcal{T}_a^c$	$\mathcal{T}_\mathcal{O}^c$

In order to adjust the **CLuNs**-semantics to $\mathcal{L}_a$, it is desirable to extend the assignment function v in such a way that it also assigns a member of D to every member of $\mathcal{T}_\mathcal{O}^c$. This is realized by clauses as the following: if $v(t) = \mathfrak{a}$ then $v(t') = \mathfrak{b}$ iff $\langle \mathfrak{a}, \mathfrak{b} \rangle \in v(')$—note that $v(')$ is a functional (one-one or many-one) binary relation. To avoid confusion, I mention the adjusted definition of $[\![A]\!]$: for all $A \in \mathcal{W}_\mathcal{O}^a$,

$\alpha, \beta \in \mathcal{T}_{\mathcal{O}}^c$, and $B \in \mathcal{W}_{\mathcal{O}}^a$, $[\![A]\!]$ is the smallest set S fulfilling: (i) $A \in S$, and (ii) if $B(\alpha) \in S$ and $v(\beta) = v(\alpha)$, then $B(\beta) \in S$. We also need to adjust **CLuNs**^{m} to $\mathcal{L}_a$. There are two non-obvious changes (i) $\Omega = \{\exists (t_1 = t_2 \wedge \neg t_1 = t_2) \mid t_1, t_2 \in \mathcal{T}_a\}$ and (ii) references to (2.9) and to (2.12) should be reinterpreted in that $B_1, \ldots, B_m \in \mathcal{F}_s^a$ is replaced by $B_1, \ldots, B_m \in \mathcal{T}_a$.[21]

As we shall consider models that verify inconsistencies, some terms will be called intuitively identical and others intuitively different. One might see these terms as defined by **PA**, supposing that it is consistent. This is made more precise as follows. Let $\mathcal{N}$ be the set of numerals and consider a function $f \colon \mathcal{T}_a^c \to \mathcal{N}$ that maps constant terms to the corresponding numeral. An obvious way to define f is in terms of a sequence of systematic transformations delineated by Fact 2 below. Where $t_1, t_2 \in \mathcal{T}_a^c$, t_1 and t_2 are *intuitively identical* iff $f(t_1) = f(t_2)$ and *intuitively different* otherwise. Where $t_1, t_2 \in \mathcal{T}_{\mathcal{O}}$, t_1 and t_2 are *intuitively identical* iff every 'joint instantiation' results in two intuitively identical constant terms. This means: if one systematically replaces, in t_1 as well as t_2, every member of $\mathcal{V} \cup \mathcal{O}$ by a numeral, then one obtains two members of $\mathcal{T}_a^c$ that are intuitively identical. Where $t_1, t_2 \in \mathcal{T}_{\mathcal{O}}$, t_1 and t_2 are *intuitively different* iff every 'joint instantiation' results in two intuitively different constant terms.

To avoid confusion I add two comments. (i) All $t_1, t_2 \in \mathcal{T}_a^c$ are either intuitively identical or intuitively different, but not both. Some $t_1, t_2 \in \mathcal{T}_{\mathcal{O}}$ are neither intuitively identical nor intuitively different, but none are both. (ii) That the intuitive identity [difference] of members of $\mathcal{T}_{\mathcal{O}}$ is decided in terms of instantiations with numerals does not rule out non-standard numbers; example: $x' + y$ and $y + x + 0'$ are intuitively identical for all numbers because they are for all natural numbers.[22]

Let Γ_1 comprise P1–P6, P8, and all formulas of the form P7.

P1 $\quad \forall x \neg x' = 0$
P2 $\quad \forall x \forall y (\neg x = y \supset \neg x' = y')$
P3 $\quad \forall x \, x + 0 = x$
P4 $\quad \forall x \forall y \, x + y' = (x + y)'$
P5 $\quad \forall x \, x \times 0 = 0$
P6 $\quad \forall x \forall y \, x \times y' = (x \times y) + x$
P7 $\quad A(0) \supset (\forall x (A(x) \supset A(x')) \supset \forall x A(x))$

---

[21] The modification to these statements concerning **CLuNs** is not a mistake. The modified statements are easily provable and the involved contradictions should be members of Ω, which is redefined.

[22] See for example Boolos, Burgess, and Jeffrey (2002) for details on standard and non-standard numbers.

P8 $\exists x \forall y (\neg y = y \supset y = x)$

The 'usual' version of P2 is

P2a $\forall x \forall y (x' = y' \supset x = y)$.

Consider the following sets:

Γ_0:	P1	P2a	P3	P4	P5	P6	P7	
Γ_1:	P1	P2	P3	P4	P5	P6	P7	P8
Γ_2:	P1	P2	P3	P4	P5	P6	P7	

The 'usual' set of Peano axioms will be called Γ_0 and $\mathbf{PA} = \langle \Gamma_0, \mathbf{CL} \rangle$. As suggested before, $\mathbf{APA} = \langle \Gamma_1, \mathbf{CLuNs}^m \rangle$. I shall sometimes identify $\mathbf{APA}$ with its set of theorems.

Set of $\mathbf{APA}$-theorems: $Cn_{\mathbf{CLuNs}^m}(\Gamma_1)$.

I shall often need to refer to Γ_2, which is a proper subset of Γ_1. This is related to the fact that P8 is a member of Γ_1 for a rather special reason, which will become clear in Section 5. P2 is $\mathbf{CL}$-equivalent to P2a and P8 is a $\mathbf{CL}$-theorem. So:

Fact 1 $Cn_{\mathbf{CL}}(\Gamma_1) = Cn_{\mathbf{CL}}(\Gamma_0) = \mathbf{PA}$.

So $\mathbf{APA}$ is well-defined, but in view of (3.15), it may fail to be semi-recurse. Even if $\mathbf{PA}$ is inconsistent, however, $\mathbf{APA}$ is well-defined and, as shown in Section 5, in an absolute sense non-trivial. So $\mathbf{APA}$ is better off than $\mathbf{PA}$, which cannot be shown to be non-trivial in an absolute sense. Why is $\mathbf{APA}$ defined in terms of a different set of axioms than the usual Peano axioms? This question will be answered in two steps. First, however, I present a result that will prove useful in the sequel.

Fact 2 *P3–P6 together with all of the following are $\mathbf{APA}$-theorems:*

$$\forall x \, x' = x + 0' \tag{1}$$
$$\forall x \, x \times 0' = x \tag{2}$$
$$\text{where } \odot \in \{+, \times\}, \, \forall x \forall y \, x \odot y = y \odot x \tag{3}$$
$$\text{where } \odot \in \{+, \times\}, \, \forall x \forall y \forall z (x \odot y) \odot z = x \odot (y \odot z) \tag{4}$$
$$\forall x \forall y \forall z \, (x + y) \times z = (x \times z) + (y \times z) \tag{5}$$
$$\forall x \forall y \forall z \, (x \times y) + z = (x + z) \times (y + z) \tag{6}$$

These theorems are $\mathbf{CLuNs}$-consequences of P3–P7. The $\mathbf{CLuNs}$-proofs are identical to the corresponding $\mathbf{CL}$-proofs. Mentioning P3–P6 in Fact 2 is redundant but simplifies references.

Lemma 1 *Where* **CLuNs** *is defined over* $\mathcal{L}_a$, $Cn_{\mathbf{CLuNs}}(\{\forall x \forall y\, x = y, \exists x \exists y \neg x = y\}) = \mathcal{W}_a$.

Proof. Let $M_1 = \langle D, v \rangle$ verify $\{\forall x \forall y\, x = y, \exists x \exists y \neg x = y\}$.[23] So D is a singleton and there are $\alpha, \beta \in \mathcal{C} \cup \mathcal{O}$ for which $v_{M_1}(\alpha = \beta) = v_{M_1}(\neg \alpha = \beta) = 1$. This is only possible if $v(\neg B) = 1$ for a $B \in [\![\alpha = \beta]\!]$.

Consider any $t_1 = t_2 \in \mathcal{W}_\mathcal{O}^a$. As D is a singleton, $v(t_1) = v(t_2) = v(\alpha) = v(\beta)$. It follows that $v_{M_1}(t_1 = t_2) = 1$ and that $[\![\alpha = \beta]\!] = [\![t_1 = t_2]\!]$. But then $v_{M_1}(\neg t_1 = t_2) = 1$.

So we have shown that $M_1 \Vdash A$ for all $A \in \mathcal{W}_\mathcal{O}^a$. But then $M_1 \Vdash A$ for all $A \in \mathcal{W}_a$.[24] ■

Theorem 1 *If* $\Gamma \subseteq \mathcal{W}_a$, $\mathrm{P1} \in \Gamma$, *and* $\Gamma \vdash_{\mathbf{CLuNs}^m} \forall x \forall y\, x = y$, *then* $Cn_{\mathbf{CLuNs}^m}(\Gamma)$ *is trivial.*

Proof. Suppose that the antecedent is true. As $\mathrm{P1} \in \Gamma$, $\Gamma \vdash_{\mathbf{CLuNs}} \exists x \exists y \neg x = y$. So $\forall x \forall y\, x = y, \exists x \exists y \neg x = y \in Cn_{\mathbf{CLuNs}^m}(\Gamma)$. But then $Cn_{\mathbf{CLuNs}^m}(\Gamma)$ is trivial in view of (3.2), (3.8), and Lemma 1. ■

Define $x \leq y =_{df} \exists z\, x + z = y$ and $x < y =_{df} \exists z\, x + z' = y$. Whether **PA** is consistent or not, all **CLuNs**-consequences of Γ_2 are **APA**-theorems. Below I list some of these theorems; comments follow after the list.

$$\forall x \forall y (x + y' = x' + y) \tag{7}$$

$$\forall x \forall y \neg x + y' = 0 \tag{8}$$

$$\forall x (0 = x \vee \exists y\, y' = x) \tag{9}$$

$$\forall x \forall y \exists z (x + z = y \vee y + z = x) \tag{10}$$

$$\forall x \forall y \exists z (x + z = y \vee y + z' = x) \tag{11}$$

$$\forall x \forall y (x = y \vee \exists z\, x + z' = y \vee \exists z\, y + z' = x) \tag{12}$$

$$\forall x \neg x' = x \tag{13}$$

$$\forall x \forall y \neg x + y' = x \tag{14}$$

$$\forall x \forall y \exists \dot{v} \exists \dot{w}((y' \times v) + w = x \wedge \exists z\, w + z' = y') \tag{15}$$

Here are some proof outlines, which demand a bit of work from the reader, as well as some comments on the meaning of the theorems. Most proofs require **CLuNs**-properties, especially RoI. The dot on top of some variables indicate that these are specific, receiving their meaning from an existential quantifier, and depend on the other variables.

---

[23] A concrete example, called M_3, comes up in Section 4.

[24] The reference to $\mathcal{L}_a$ and to the connected $\mathcal{W}_\mathcal{O}^a$ is essential. If the language is extended, for example, with a unary predicate P, extending M_1 to the new language may result in $M_1 \not\Vdash \forall x \neg Px$ (or else in $M_1 \not\Vdash \forall x Px$). So triviality is easily avoided for that language.

(7): by Fact 2.

(8): by P1, Fact 2, and P7.

(9): by P2 and P7 (obvious induction over x). Every number is either zero or has a predecessor.

(10): by (9), P2, P7, and Fact 2; $\forall y \exists z (0 + z = y \lor y + z = 0)$ is the basis of the induction. Of every two numbers, one can be reached by adding a number to the other.

(11), (12): similar to (10). Between two numbers holds $\leq$ or $>$. Between two numbers holds $=$ or $<$ or $>$.

(13), (14): by P1, P2, and P7. Every number is different from its successor. Every number is different from every larger number.

(15): by an induction over x. The basis is $\forall y \exists v \exists w ((y' \times v) + w = 0' \land \exists z\, w + z' = y')$, which holds by Fact 2, setting $v = 1$, $w = 0$, and $z = y$ if $y = 0$ and setting $v = 0$, $w = 0'$, and $z' = y$ otherwise—the latter is possible in view of (9). Suppose $(y' \times \dot{v}) + \dot{w} = x \land \exists z\, \dot{w} + z' = y'$; so $(y' \times \dot{v}) + \dot{w}' = x'$ in view of Fact 2; if $\exists z\, \dot{w}' + z' = y'$ is false, then $\exists z\, y' + z = \dot{w}'$ by (11); let $\dot{u}$ be this z; so $(y' \times \dot{v}') + \dot{u} = x'$ and $\exists z\, \dot{u}' + z' = y'$ by Fact 2. For all x and y there is a v and there is a w such that x divided by y' is v remainder w (Euclidean division).

Let us now discuss the reason for replacing P2a by P2. Suppose that $\exists x\, x' = x \in Cn_{\mathbf{CLuNs}^m}(\Gamma_0)$ and consider the following **CLuNs**-proof. I do not list a full justification but point out that the formula is a **CLuNs**-theorem or make the derivation sufficiently clear by referring to previous lines, to members of Γ_0 and to derivable rules of **CLuNs**. The rule in the justification of line 6 requires that x is not free in B.

1	$\exists x\, x' = x$	Premise
2	$\forall x(x'' = x' \supset x' = x)$	P2a
3	$\forall x((x' = x \supset 0' = 0) \supset (x'' = x' \supset 0' = 0))$	2; $A \supset B / (B \supset C) \supset (A \supset C)$
4	$0' = 0 \supset 0' = 0$	**CLuNs**-theorem
5	$\forall x(x' = x \supset 0' = 0)$	4, 3; P7
6	$0' = 0$	5, 1; $\forall x(A(x) \supset B), \exists x A(x) / B$
7	$\forall x(x + 0' = x + 0)$	6; $a = b / \forall x(x + a = x + b)$
8	$\forall x(x' = x)$	7; by P3 and P4
9	$0 = 0$	**CLuNs**-theorem
10	$\forall x(x = 0 \supset x' = 0)$	8; $f(x) = x / A(x) \supset A(f(x))$
11	$\forall x\, x = 0$	9, 10; P7
12	$\forall x \forall y\, x = y$	11; $x = a, y = a / x = y$

So $\exists x\, x' = x \in Cn_{\mathbf{CLuNs}^m}(\Gamma_0)$ entails $\forall x \forall y\, x = y \in Cn_{\mathbf{CLuNs}^m}(\Gamma_0)$ in view of (3.8). But then it also entails that $Cn_{\mathbf{CLuNs}^m}(\Gamma_0)$ is trivial in view of Theorem 1.

The unpalatable property of $Cn_{\mathbf{CLuNs}^m}(\Gamma_0)$ is avoided when P2a is replaced by P2. It is instructive to compare the following two **CLuNs**^{m}-

proofs.

1	$18 = 17$	Premise	$\emptyset$
2	$18 = 17 \supset 17 = 16$	P2a; RU	$\emptyset$
3	$17 = 16$	1, 2; RU	$\emptyset$

1	$18 = 17$	Premise	$\emptyset$
2	$\neg 17 = 16 \supset \neg 18 = 17$	P2	$\emptyset$
3	$17 = 16$	1, 2; MT	$\{(18 = 17) \wedge \neg(18 = 17)\}$ ✓[40]
4	$\neg 1 = 0$	P1	$\emptyset$
5	$\neg 1 = 0 \supset \neg 2 = 1$	P2	$\emptyset$
6	$\neg 2 = 1$	4, 5; MP	$\emptyset$
7	$\neg 2 = 1 \supset \neg 3 = 2$	P2	$\emptyset$
8	$\neg 3 = 2$	6, 7; MP	$\emptyset$
⋮			
39	$\neg 18 = 17$	37,38; MP	$\emptyset$
40	$18 = 17 \wedge \neg 18 = 17$	1, 39; $A, B/A \wedge B$	$\emptyset$

So the central difference between both premise sets depends essentially on (2.2) and (2.3). If the **CLuNs**^{m}-consequence set of Γ_0 contains a statement to the effect that a number is its own successor, then it follows that each *smaller* number is also its own successor. It also follows by **CL**$^+$ that each *larger* number is its own successor. More importantly, one easily establishes that all numbers are their own successor (from $0' = 0$ and the **CLuNs**-theorem $\forall x(x' = x \supset x'' = x')$ by P7). The situation for Γ_1 is wholly different. If the **CLuNs**^{m}-consequence set of Γ_1 contains a statement to the effect that a certain number n is its own successor, then it still **CLuNs**^{m}-follows that all larger numbers are their own successor, but not that any smaller number is its own successor.[25] So, according to **APA**, a number can be identical to its successor without being identical to its predecessor. The same difference surfaces if $18 = 17$ is replaced by $\exists x\, x' = x$—one then needs P7 to prove $0' = 0$. In the next section I shall show that even if $\exists x\, x' = x$ is a theorem of **APA**, then **APA** is non-trivial and remarkably close to **PA**.

Fact 2 guarantees that terms are identical [different] in **APA** whenever they are intuitively identical [different]. Of course, the terms may *also* be different [identical]. For example, some models of Γ_1 verify $3 + 4 = 4 + 3$ as well as $\neg 3 + 4 = 4 + 3$, but this is all right because they also verify $7 = 7$ as well as $\neg 7 = 7$, and $7 = 3 + 4$ as well as

---

[25]One can show that there is at least one $\varphi \in \Phi(\Gamma_1)$ such that every condition on which it can be derived that a smaller number is its own successor overlaps with φ. It would take a while to explain this, whence I ask the reader to believe me or to study the minimal abnormality strategy, for example in one of Batens (2007, 201x).

$\neg 7 = 3 + 4$, and so on. Fact 2 also warrants the provability of the following theorem—its effect becomes clear later on.

Theorem 2 *For all $t_1, t_2 \in \mathcal{T}_a$, every model of Γ_1 verifies $\forall t_1 = t_2$ $[\forall \neg t_1 = t_2]$ if t_1 and t_2 are intuitively identical [different].*

Proof. If $t_1, t_2 \in \mathcal{T}_a^c$ are intuitively identical, $t_1 = t_2$ is a **CLuNs**-consequence of Γ_1 in view of Fact 2. If they are intuitively different, then, in view of Fact 2, the two terms are respectively identical to two different numerals, say $0^{(n)}$ and $0^{(m)}$ with $n \neq m$. However, if $n \neq m$, then $\neg 0^{(n)} = 0^{(m)}$ is a **CLuNs**-consequence of P1 and P2; hence $\neg t_1 = t_2$ is a **CLuNs**-consequence of Γ_1.

A proof that every 'joint instantiation' in $t_1, t_2 \in \mathcal{T}_\mathcal{O}$ results in two intuitively identical [different] numbers, can obviously be transformed to a **CLuNs**-proof of $\forall t_1 = t_2$ $[\forall \neg t_1 = t_2]$ in view of Fact 2.[26] ∎

The most handy intuitive criterion for the intuitive difference between two numbers x and y is that $\exists z(x + z' = y \lor y + z' = x)$ is **CLuNs**-derivable from Γ_2 together with what is given about x and y.

The reason for including P8 in Γ_1 will become clear in Section 5. In preparation, I shall pay attention to both Γ_1 and Γ_2.

Let us now consider some models that verify Γ_2. These have the following successor graphs, among others:[27]

type 1: $\mathfrak{f} \to \mathfrak{g} \to \cdots$

type 2: $\mathfrak{f} \to \mathfrak{g} \to \cdots \to \mathfrak{a} \circlearrowleft$

type 3: $\mathfrak{f} \to \mathfrak{g} \to \cdots \to \mathfrak{a} \to \mathfrak{g}$ (cycle)

type 4: $\mathfrak{f} \to \cdots \to \mathfrak{h} \to \mathfrak{g} \to \cdots \to \mathfrak{a} \to \mathfrak{g}$ (tail into cycle)

---

[26] From the metatheoretic proof follows that there is a proof of $\Gamma_1 \vdash_{\textbf{CLuNs}} f([\neg]t_1 = t_2)$ in which $f([\neg]t_1 = t_2)$ results from instantiating $[\neg]t_1 = t_2$ with dummy constants; next one applies Universal Generalization. P7 will be required in the transformed proof iff it is present in the original one.

[27] The graphs of the types 2–4 occur in the context of finite inconsistent models of arithmetic in Priest (1997).

The understanding is that 0 names $\mathfrak{f}$, that the arrow represents the successor function ($v(\alpha') = \mathfrak{x}$ iff there is an arrow from $v(\alpha)$ to $\mathfrak{x}$), and that addition and multiplication are calculated in terms of the successor function. Identity and difference of constant terms is as determined by the assignment function v (see C= and C¬).

If **PA** is consistent, Γ_2 has models of all these types. Type 1 models are at least countably infinite. The standard model is a type 1 model and so are the non-standard models. On the usual understanding, these models are consistent. In other words, they are equivalent to models in which $v(\neg A) = 0$ for all $A \in \mathcal{W}_\mathcal{O}^a$. However, some type 1 models of Γ_2 are inconsistent, for example because they verify $\neg t = t$ for some $t \in \mathcal{T}_\mathcal{O}$. In order to be a model of Γ_2, the inconsistent type 1 models should also verify $\neg t' = t'$, $\neg t'' = t''$, etc., in view of P2. They need not, however, verify any other inconsistency. From a certain point on, all numbers are stated to be different from themselves, but they need not be identical to any number that is intuitively different from them.

In type 2 models, $\mathfrak{a}$ may be at any distance from $\mathfrak{f}$. Finite type 2 models were invoked by Van Bendegem (1993) in order to turn his finitist arithmetic into an inconsistent finitist arithmetic that extends classical arithmetic. The models were further studied in Priest (1994); Van Bendegem (1994). In a type 1 model, the number of arrows between $\mathfrak{f}$ and $\mathfrak{a}$ determines which numeral, say $0^{(n)}$, refers to $\mathfrak{a}$. The same holds in type 2 models. However, $0^{(n+1)}$ *also* refers to $\mathfrak{a}$ because of the arrow from $\mathfrak{a}$ to $\mathfrak{a}$. So, all greater numerals also refer to $\mathfrak{a}$ because one may continue to go from $\mathfrak{a}$ to $\mathfrak{a}$. It follows that $0^{(n+1)} = 0^{(n)}$. However, in view of P1 and P2, we also need $v_M(\neg 0^{(n+1)} = 0^{(n)}) = 1$ and hence also $v_M(\neg 0^{(n)} = 0^{(n)}) = 1$.[28] And this goes on: $0^{(n+m)} = 0^{(n)}$ is verified along with $\neg 0^{(n+m)} = 0^{(n)}$ and hence also $\neg 0^{(n+m)} = 0^{(n+m)}$.

Type 3 models are circular: $\mathfrak{f}$, which functions as the number zero and is named by 0, is the successor of another number, say named by $t \in \mathcal{T}_\mathcal{O}^c$. So the model verifies $t' = 0$, but it should also verify $\neg t' = 0$ in view of P1. It follows that type 3 models of Γ_2 verify $0 = 0 \wedge \neg 0 = 0$. The inconsistencies spread upwards (by P2 and **CLuNs**) and spread to constant terms: the model for example verifies $\neg 0'' = 0' + 0'$. This does not mean that the model is trivial; if its cardinality is larger than 1; it falsifies $0' = 0$, $\exists x(x' = x)$, etc.

Type 4 models resemble type 3 models, except that they have a tail in front of the cycle and that zero is located at the tip of the tail. Note that the number $\mathfrak{h}$ has two predecessors, one in the tail and one in the cycle. The latter number is larger than $\mathfrak{h}$ and so causes an inconsistency

---

[28] So there are $t_1, t_2 \in \mathcal{T}_\mathcal{O}^c$ for which $v(t_1) = v(t_2) = \mathfrak{a}$ while $v(\neg t_1 = t_1) = 1$.

at the level of the language. Let $t_1 \in \mathcal{T}_{\mathcal{O}}^c$ name $\mathfrak{a}$ and let $t_2 \in \mathcal{T}_{\mathcal{O}}^c$ name $\mathfrak{h}$. We know from the graph that the model verifies $t_1' = t_2$ and as it is a model of Γ_2, it also verifies $\neg t_1' = t_2$ (in view of P1, P2, and P7). As is the case for type 3 models, type 4 models verify $t = t \wedge \neg t = t$ whenever $t \in \mathcal{T}_{\mathcal{O}}^c$ names a number in the cycle ($\mathfrak{h}$ included), but unlike what is the case in type 3 models, some numbers are not in the cycle but in the tail.

The graphs of types 2–4 occur in Priest (1997) as graphs of finite inconsistent models of arithmetic (the supposedly consistent set of formulas verified by the standard model of arithmetic). This concerns **LP**-models, in which there is no detachable implication. However, there are also infinite models of Γ_2 that have those graphs. Moreover, once there is a cycle in a model of Γ_2, nothing comes after it. Indeed, let a refer to the object with which the cycle starts and let b refer to the number of objects in the cycle—$b = 1$ in type 2 models and $b \geq 1$ in type 3 and type 4 models. Suppose that there is a number c that is larger than a: $\exists x\, a + x' = c$ and let $x' = d$. In view of (15), $\exists v \exists w ((b \times v) + w = d \wedge \exists z\, w + z' = b)$. But then $b = a \dotplus w$, which is an object in the cycle. So although (15) does not exclude that a cycle contains several non-standard blocks, it warrants that no number is consistently larger[29] than all numbers in a cycle.[30] Obviously, all numbers in the cycle are larger as well as smaller than each other and larger as well as smaller than themselves.

In preparation of the next section, let me explain in which way models of types 2–4 may be transformed to less abnormal models. A type 4 model may be seen as a rope with knots, most of which is wound on the cylinder of a windlass. The knots line up on the cylinder. If the circumference of the cylinder is, for example, 4 knots, then the knot named by $0^{(n)}$ lines up with the knot named by $0^{(n+4)}$, and so on. Knots that line up with each other are identified in that term referring to the one also refers to the other. So, in the 4 knots example, the model will verify $0^{(n)} = 0^{(n+4)}$ because both names refer to the same entity, which is a

---

[29] In a given model a number is consistently larger than another iff it is larger and not also not larger (identical or smaller). We shall see in Section 6 that it can be expressed within **APA** that a member of $\mathcal{W}_a$ is consistently true. We do not need that, however, for the point made in the text.

[30] It follows that the infinite **LP**-models of arithmetic from Priest (2000) are not **APA**-models. The **LP**-models verify P7 because, in **LP**, P7 reduces to $\neg A(0) \vee (\exists x(A(x) \wedge \neg A(x')) \vee \forall x A(x))$, which is true as soon as one of the three disjuncts is true. The relevant instance of this is true because the **LP**-model verifies not only $a + (0 \times b) = a$ but also its negation, which corresponds to $\neg A(0)$ in the schema. The situation becomes different when **LP** is extended with an implication that is detachable and can be nested and when mathematical induction is formulated in terms of this implication.

set of knots lined up with each other on the cylinder. However, being a model of Γ_2, the model will verify $\neg 0^{(n)} = 0^{(n+4)}$. But then it also verifies $\neg t = t$ whenever t names a knot on the cylinder. The knots that are not on the cylinder do not line up with any other knots. So if t refers to one of them, there is no need why the model should verify $\neg t = t$. Moreover, if one has a type 4 model of Γ_2 and unwinds the rope from the cylinder until a knot comes off (and sets $v(\neg t = t) = 0$ when t names that knot), the resulting model is less abnormal than the original. Where $0^{(n)}$ was the lowest numeral that named this knot in the original model, the resulting model falsifies $\neg 0^{(n)} = 0^{(n)}$ whereas the original model verified it.

A type 2 model is a type 4 model the cylinder of which has a circumference of one knot. Transforming it to a less abnormal model proceeds as for all type 4 models. A type 3 model is a type 4 model with the whole rope wound on the cylinder. A less abnormal model is again obtained by unwinding the rope (and adjusting v); the result is a *proper* type 4 model—one that is not also a type 3 model or a type 2 model.

For all we know at this point in the paper, Γ_1 may obviously have other types of inconsistent models. We shall see, however, that all **APA**-models are of a type mentioned so far.

5 Some Results

Let us first consider some **APA**-theorems that are all **CLuNs**-consequences of Γ_1. Below I list some of those theorems and next comment on them. The theorems may not be very interesting from a mathematical point of view, but they are interesting for the present paper. Let ℓ be defined contextually: $A(\ell)$ iff $\exists x (\neg x = x \wedge \forall y (\neg y = y \supset y = x) \wedge A(x))$.[31] If $[\ell]$ occurs in a formula, it limits the scope of the definite description; otherwise the scope is the whole formula. Thus $\neg[\ell]A(\ell)$ abbreviates $\neg \exists x (\neg x = x \wedge \forall y (\neg y = y \supset y = x) \wedge A(x))$, whereas $\neg A(\ell)$ abbreviates $\exists x (\neg x = x \wedge \forall y (\neg y = y \supset y = x) \wedge \neg A(x))$. In view of (22)–(24), $x = \ell$ entails $\neg x = x$, $x' = x$, and $\forall y\, x + y = x$.

---

[31] Needless to say, the definiens requires relettering if x is not free in $A(x)$.

$$\forall x \forall y ((x = y \land \neg x = y) \supset \neg x = x) \tag{16}$$
$$\forall x (\neg x = x \equiv x' = x) \tag{17}$$
$$\forall x (x' = x \equiv \exists y\, x + y' = x) \tag{18}$$
$$\forall x (x' = x \equiv \forall y\, x + y = x) \tag{19}$$
$$\forall x (x' = x \equiv \forall y \forall z (x + y = z \supset z = x)) \tag{20}$$
$$\forall x (\neg x = x \supset [\ell] x = \ell) \tag{21}$$
$$\exists x \neg x = x \supset [\ell] \neg \ell = \ell \tag{22}$$
$$\exists x \neg x = x \supset [\ell] \ell' = \ell \tag{23}$$
$$\exists x \neg x = x \supset [\ell] \forall x\, \ell + x = \ell \tag{24}$$
$$\exists x \neg x = x \supset [\ell] \forall x \neg x = \ell \tag{25}$$
$$\forall x \forall y (x + y' = x \supset [\ell] x = \ell) \tag{26}$$
$$\forall x \forall y (x' = y' \supset (x = y \lor [\ell] x' = \ell)) \tag{27}$$
$$\forall x \forall y \forall z \forall w ((x + z' = y \land y + w = x) \supset [\ell](x = \ell \land y = \ell)) \tag{28}$$
$$\forall x \forall y (x' = y' \supset (x = y \lor [\ell] x = \ell \lor [\ell] y = \ell)) \tag{29}$$
$$\forall x \forall y (x' = y' \supset (x = y \lor x' = x \lor y' = y)) \tag{30}$$

Here are some proof outlines—they demand a bit of work from the reader—as well as some comments on the meaning of the theorems. A dot on top of a variable is used as before.

(16): suppose $x = y$ and $\neg x = y$; so $\neg x = x$ by RoI.

(17): $\Rightarrow$ suppose $\neg x = x$; so $\neg x' = x'$ by P2; but then $x' = x$ by P8; $\Leftarrow$ suppose $x' = x$; $\neg x' = x$ by (13); so $\neg x = x$ by RoI.

(18): similar to (17) relying on (14).

(19): $\Rightarrow$ suppose $x' = x$; so $x' + y = x + y$ as well as $x + y' = x + y$ by Fact 2; hence also $\forall y(x + y = x \supset x + y' = x)$; from this and $x + 0 = x$, which holds in view of P3, $\forall y(x + y = x)$; $\Leftarrow$ suppose $x + y = x$; so $\neg x = x$ by (14) and $x' = x$ by (17). A number is its own successor iff it is identical to all larger numbers.

(20): as (19), to which this is a variant.

(21), (22): by P8.

(23): suppose $\exists x \neg x = x$; by (22) $\neg \ell = \ell$; by P2 $\neg \ell' = \ell'$; by (21) $\ell' = \ell$.

(24): from (23), P7, and Fact 2.

(25): suppose $x = \ell$; so $\neg x = \ell$ by (22); but $\vdash_{\mathbf{CLuNs}} (A \supset \neg A) \supset \neg A$. If a number is different from itself, which holds just in case **PA** is inconsistent, then the number ℓ exists and is different from itself as well as from all other numbers—the latter are consistently different from ℓ.

(26): suppose $x+y' = x$; $\neg x+y' = x$ by (14); from both $\neg x = x$; so $x = \ell$ by (21). Any number identical to an intuitively larger number is ℓ.

(27): suppose $x' = y'$; $\vdash_{\mathbf{CLuNs}} x = y \vee \neg x = y$; suppose $\neg x = y$; so $\neg x' = y'$ by P2; this together with $x' = y'$ **CLuNs**-entails $\neg x' = x'$; hence $x' = \ell$ by (21). If a number has two consistently different predecessors, it is ℓ.[32]

(28): by (26). If $x < y$ and $y \leq x$, then $x = y = \ell$.

(29): suppose $x' = y'$; by (27) $x = y$ or $x' = \ell$; suppose $x' = \ell$; by (10) $x+\dot{z} = y \vee y+\dot{z} = x$; (i) suppose $x+\dot{z} = y$; (i.i) suppose $\dot{z}$ is 0; then $x = y$ by P3; (i.ii) suppose $\dot{z}$ is larger than 0: by (9) $\dot{z}$ has a predecessor, say $\dot{w}$; by (i) $x + \dot{w}' = y$, whence $x' + \dot{w} = y$ by (7); as $x' = \ell$, $y = \ell$ by (24) and Fact 2. (ii) suppose $y + \dot{z} = x$; by the same reasoning either $x = y$ or $x = \ell$. If a number has two consistently different predecessors, then (the number is ℓ by (27) and moreover) one of these predecessors is ℓ. Put differently, no number has two predecessors that are consistently different from this number as well as from each other.

(30): from (29).

The importance of (29) is that it rules out 'branching to the past', viz. models that have a successor graph of the following form—I gave $\mathfrak{c}_1$ some predecessors in view of (9).

$$\ldots \to \mathfrak{c}_1 \searrow$$
$$\mathfrak{f} \to \ldots \to \mathfrak{a}_2 \to \mathfrak{a}_1 \to \mathfrak{b} \circlearrowright$$

Lemma 2 $Cn_{\mathbf{CLuNs}}(\Gamma_1)$ *is non-trivial.*

Proof. Consider a finite model $M_2 = \langle D, v \rangle$ with the following properties: $D = \{\mathfrak{f}, \mathfrak{a}\}$; the interpretation of successor, addition and multiplication are as specified by the tables below; $v(0) = \mathfrak{f}$, and $v(\neg A) = 1$ iff A is $0' = 0'$.

$\mathfrak{f} \to \mathfrak{a} \circlearrowright$

	$'$
$\mathfrak{f}$	$\mathfrak{a}$
$\mathfrak{a}$	$\mathfrak{a}$

$+$	$\mathfrak{f}$	$\mathfrak{a}$
$\mathfrak{f}$	$\mathfrak{f}$	$\mathfrak{a}$
$\mathfrak{a}$	$\mathfrak{a}$	$\mathfrak{a}$

$\times$	$\mathfrak{f}$	$\mathfrak{a}$
$\mathfrak{f}$	$\mathfrak{f}$	$\mathfrak{f}$
$\mathfrak{a}$	$\mathfrak{f}$	$\mathfrak{a}$

---

[32]Suppose that **PA** is consistent. As will appear in the sequel, the following then hold in **APA**: (i) $x' = \ell$ is consistently false for all x and this can be stated by an $A \in \mathcal{W}_a$, and (ii) P2a is a **CLuNs**-consequence of this A together with (27).

As $v(0) = \mathfrak{f}$, $\mathfrak{a} = v(0') = v(0'') = \ldots$—so every member of the domain of M_2 is named by a member of $\mathcal{T}_a$, whence we can disregard members of $\mathcal{T}_\mathcal{O} - \mathcal{T}_a$. Universally quantified sentences reduce to their finite number of instances. For example, M_2 falsifies both $0' = 0$ and $0'' = 0$. So it verifies $\neg 0' = 0$ and $\neg 0'' = 0$, and hence also P1. That M_2 verifies all other members of Γ_1 is shown similarly. As M_2 falsifies $0' = 0$, it is a non-trivial model of Γ_1.[33] ∎

Instead of a type 2 model with cardinality 2, I could have chosen a type 2 model with any finite cardinality larger than 1. The proof would become lengthier, but no new difficulty would arise. The reader can easily extend this result to type 3 models and to proper type 4 models, except that these only verify Γ_2. This results in the following fact.

Fact 3 *For every $n > 1$, Γ_1 has a non-trivial type 2 model of cardinality n, Γ_2 has a non-trivial type 3 model of cardinality n, and Γ_2 has a non-trivial proper type 4 model of cardinality $n + 1$.*[34]

Fact 4 *For every $n > 1$, Γ_1 has a non-trivial type 2 model M of cardinality n such that, for all $t_1, t_2 \in \mathcal{T}_a^c$, $M \Vdash t_1 = t_2 \wedge \neg t_1 = t_2$ iff both $v(t_1)$ and $v(t_2)$ refer to the number that is its own successor.*

For every $n > 1$, Γ_2 has a non-trivial type 3 model M of cardinality n such that, for all $t \in \mathcal{T}_a^c$, $M \Vdash t = t \wedge \neg t = t$.

For every $n > 2$, Γ_2 has a non-trivial type 4 model M of cardinality n such that, for all $t \in \mathcal{T}_a^c$, $M \Vdash t = t \wedge \neg t = t$ iff $v(t)$ refers to an element in the cycle.

Theorem 3 APA *is non-trivial.*

Proof. From Lemma 2 in view of (3.7). ∎

The reader may wonder what we win by going adaptive: Which are the sentences in $\{A \mid \vdash_{\mathbf{APA}} A\} - Cn_{\mathbf{CLuNs}}(\Gamma_1)$? I shall answer this question later, but already illustrate here that the set is not empty. We know from Lemma 2 that M_2 is a model of Γ_1 and that M_2 falsifies $0' = 0$ and verifies $\neg 0' = 0$. This obviously does not mean, that every model of Γ_1 falsifies $0' = 0$. Γ_1 has other models and some of them verify sentences falsified by M_2. An example is M_3: $D = \{\mathfrak{a}\}$ and the successor graph is the one to the left below. There is not much choice for interpreting the language: $\mathfrak{a}' = \mathfrak{a} + \mathfrak{a} = \mathfrak{a} \times \mathfrak{a} = \mathfrak{a}$, $v(0) = \mathfrak{a}$, and $v(\neg A) = 1$ iff A is $0 = 0$.

---

[33] M_2 falsifies P2a and hence Γ_0. It verifies infinitely many inconsistencies, for example, for all $n > 0$, $0^{(n)} = 0^{(n)} \wedge \neg 0^{(n)} = 0^{(n)}$ and $0^{(n+1)} = 0^{(n)} \wedge \neg 0^{(n+1)} = 0^{(n)}$.

[34] The simplest proper type 4 model has this successor graph: $\mathfrak{f} \to \mathfrak{a} \rightleftarrows \mathfrak{b}$.

$$\mathfrak{a} \circlearrowleft \qquad\qquad \mathfrak{f} \to \mathfrak{m} \to \mathfrak{a} \circlearrowleft$$

As M_3 verifies $\forall x \forall y\, x = y$ as well as $\exists x \exists y \neg x = y$, M_3 is a trivial model in view of Theorem 1. Model M_2 to the contrary falsifies $0' = 0$ as well as $\exists x\, x' = 0$. Clearly $Ab(M_2) \subset Ab(M_3)$; for example, $0' = 0 \land \neg 0' = 0 \in Ab(M_3) - Ab(M_2)$. So M_3 is not a minimal abnormal model of Γ_1. This does not clarify much, however, because eliminating the trivial model from a semantics (or adding the trivial model to it) does not influence the consequence relation. So let me take another example: the model M_4 that verifies Γ_1 and has the successor graph displayed above at the right hand side. $D = \{\mathfrak{f}, \mathfrak{m}, \mathfrak{a}\}$, $v(0) = \mathfrak{f}$, v maps every member of $\mathcal{O}$ to D, $\langle \mathfrak{x}, \mathfrak{y} \rangle \in v(')$ iff there is an arrow from $\mathfrak{x}$ to $\mathfrak{y}$ in the graph, $\mathfrak{x} + \mathfrak{y} = \mathfrak{f}^{(\mathfrak{x}+\mathfrak{y})}$, $\mathfrak{x} \times \mathfrak{y} = \mathfrak{f}^{(\mathfrak{x} \times \mathfrak{y})}$,[35] and $v(\neg A) = 1$ iff A is $0'' = 0''$. It is easily seen that M_4 is a model of Γ_1.

Clearly $Ab(M_4) \subset Ab(M_2)$. For those in doubt: both models verify $\exists x(x' = x \land \neg x' = x)$ as well as $\exists x(x = x \land \neg x = x)$; both also verify $0^{(n+1)} = 0^{(n)} \land \neg 0^{(n+1)} = 0^{(n)}$ as well as $0^{(n)} = 0^{(n)} \land \neg 0^{(n)} = 0^{(n)}$ for $n \in \{2, 3, \ldots\}$; however M_2 moreover verifies $0'' = 0' \land \neg 0'' = 0'$ as well as $0' = 0 \land \neg 0' = 0$ whereas M_4 falsifies both of these abnormalities. One of the formulas verified by M_4 and falsified by M_2 is

$$\forall y(0' = y' \supset 0 = y) \tag{31}$$

which is an instance of P2a. In words, it says that 0 is the only number of which 1 is the successor. As $\exists x\, x' = 0$ is falsified by both models, the gain made by M_4 on M_2 may be expressed by a bounded quantifier: $\forall x \leq 1 \forall y \leq 1\,(x' = y' \supset x = y)$. As type 2 models with a larger cardinality are considered, more abnormalities will be falsified and the bound in the quantifiers will go up.

The results on finite type 2 models enable one to identify a set CID of constant identities and differences that have some interesting properties. Where μ is a set of models, M is *a least model in* μ iff $M \in \mu$ and there is no $M_1 \in \mu$ such that $Ab(M_1) \subset Ab(M)$. Remember that members of $\mathcal{T}_a^c$ name natural numbers; let $\max(t_1, t_2)$ be the largest of the numbers named by $t_1, t_2 \in \mathcal{T}_a^c$.

Definition 3 $A \in$ CID *iff (i) A has the form $t_1 = t_2$ or $\neg t_1 = t_2$, (ii) $t_1, t_2 \in \mathcal{T}_a^c$, and (iii) the least models in the set of type 2 models of Γ_1 that have $\max(t_1, t_2) + 1$ as their cardinality verify A.*

---

[35]Superscripted expressions in this statement in the text are meant in the naive sense. The member of D denoted by, for example, $\mathfrak{f}^{(\mathfrak{x}+\mathfrak{y})}$ is obtained by, starting from $\mathfrak{f}$, first taking $\mathfrak{x}$ successor steps forward and next taking $\mathfrak{y}$ successor steps forward.

Fact 5 *If $t_1, t_2 \in \mathcal{T}_a^c$, then exactly one of $\{t_1 = t_2, \neg t_1 = t_2\}$ is a member of* CID.

Some type 2 models of Γ_1 verify the negation of members of CID. However, as the reader may easily check, (i) all those models also verify the members themselves and (ii) for every $A \in$ CID, those models falsify $\neg A$ whenever their cardinality is larger than the maximum referred to in Definition 3.

Fact 6 *If $A \in$ CID and M is a type 2 model of Γ_1, then $M \Vdash A$.*

Fact 7 *If $t_1, t_2 \in \mathcal{T}_a^c$, and $A \in \{t_1 = t_2, \neg t_1 = t_2\} \cap$ CID, then $M \nVdash \neg A$ holds for every type 2 model M of Γ_1 that has the cardinality $\max(t_1', t_2')$ or larger.*

Fact 8 *If $t_1, t_2 \in \mathcal{T}_a^c$ are intuitively identical [different], then $t_1 = t_2 \in$ CID $[\neg t_1 = t_2 \in$ CID$]$.*

What we can say about the further properties of **APA** will depend on the answer to certain questions. Some of the answers we do not know and possibly will never know, for example whether **PA** is consistent or inconsistent. So we have to consider several possibilities.

Lemma 3 **APA** *is consistent iff* **PA** *is consistent.*

Proof. In view of Fact 1, **PA** is consistent iff $Cn_{\mathbf{CL}}(\Gamma_1)$ is consistent. In view of (2.8), $Cn_{\mathbf{CL}}(\Gamma_1)$ is consistent iff **APA** is consistent. ∎

Theorem 4 *If **PA** is consistent, $\{A \in \mathcal{W}_a \mid \vdash_{\mathbf{APA}} A\} = \{A \in \mathcal{W}_a \mid \vdash_{\mathbf{PA}} A\}$ and vice versa. (**APA** and **PA** have the same theorems iff **PA** is consistent.)*

Proof. ⇒ Suppose that **PA** is consistent. By Lemma 3, **APA** is consistent. So, in view of (3.3), every **PA**-theorem is an **APA**-theorem and vice versa.[36]
⇐ Suppose that $\{A \in \mathcal{W}_a \mid \vdash_{\mathbf{APA}} A\} = \{A \in \mathcal{W}_a \mid \vdash_{\mathbf{PA}} A\}$. As the set of **APA**-theorems is non-trivial by Theorem 3, it follows that **PA** is consistent. ∎

Put differently, if **PA** is consistent, we loose nothing in going adaptive. Given the supposition, **PA** and **APA** have the same theorems and so have the same models, the standard model and the non-standard

---

[36] The two theories are still different, for example they have a different underlying logic.

ones. Officially, **APA**-proofs and **PA**-proofs are different because each line of an **APA**-proof has a condition as its fifth element. Moreover, the computational complexity of some **APA**-consequence sets is much higher than that of **PA**-consequence sets. As I shall show in Section 6, however, all this has no practical effects if **PA** is consistent.

Let us turn to the supposition that **PA**, and hence also **APA**, is inconsistent. In view of (2.9), it follows that a finite disjunction of abnormalities is **CLuNs**-derivable from Γ_1. I shall distinguish several possibilities, depending on the kinds of inconsistencies that can be derived.

Before considering the cases, I need to introduce the distinction between abnormalities that *affect standard numbers* and other abnormalities. I shall say that an abnormality affects a standard number iff it entails an abnormality in which all terms refer to standard numbers. Thus $0''' = 0'' \wedge \neg 0''' = 0''$ and $0^{(n)} = 0^{(n+1)} \wedge \neg 0^{(n)} = 0^{(n+1)}$ obviously affect standard numbers. Where a refers to a non-standard number, $0''' + a = 0''' + a' \wedge \neg 0''' + a = 0''' + a'$ does not affect a standard number, but $0''' + a = 0''' \wedge \neg 0''' + a = 0'''$ does because it **CLuNs**-entails $0''' = 0''' \wedge \neg 0''' = 0'''$. (The a in the previous sentence is obviously not a symbol of $\mathcal{L}_a$ but a dummy name.) Some quantified abnormalities affect standard numbers, for example $\exists x(0''' + x = 0''' \wedge \neg 0''' + x = 0''')$; indeed, whatever x refers to, it **CLuNs**-entails $0''' = 0''' \wedge \neg 0''' = 0'''$. An example of a quantified abnormality that does not affect standard numbers is $\exists x \exists y(x + 0''' = y \wedge \neg x + 0''' = y)$.[37]

Case 1: Curry-like Let us start with the worst case scenario, which is that a Curry-like paradox is finally **CLuNs**^{m}-derivable from Γ_1. If this were so, **APA** would be trivial. But we know from Theorem 3 that it is not trivial.

Case 2: Abnormalities affecting standard numbers only Suppose that a disjunction of such abnormalities is an **APA**-theorem. In the context of $\mathcal{L}_a$, the proof of (2.9) boils down to the existence of (not necessarily different) $t_1, t_2, \ldots, t_{2n-1}, t_{2n} \in \mathcal{T}_a^c$ such that

$$(t_1 = t_2 \wedge \neg t_1 = t_2) \vee \ldots \vee (t_{2n-1} = t_{2n} \wedge \neg t_{2n-1} = t_{2n}) \qquad (32)$$

is an **APA**-theorem. Each of these constant terms is intuitively identical to a specific numeral. Let m be the largest of these numerals. In view of

---

[37] A different criterion for deciding whether an abnormality affects standard numbers is presented just before Fact 11 in Section 6.

Fact 4—and also in view of Facts 6 and 7—there is a type 2 **APA**-model that falsifies (32).[38] So the main supposition is false, which gives us:

Theorem 5 *No disjunction of (one or more) abnormalities affecting standard numbers is an **APA**-theorem.*

Case 3: The General Case If **PA** is inconsistent, then, in view of Lemma 3 and (2.9), at least one **APA**-theorem is a disjunction of abnormalities:

$$\exists (A_1 \wedge \neg A_1) \vee \ldots \vee \exists (A_n \wedge \neg A_n) \vee \exists (B_1 \wedge \neg B_1) \vee \ldots \vee \exists (B_m \wedge \neg B_m) \quad (33)$$

in which each $\exists (B_i \wedge \neg B_i)$ is an abnormality affecting standard numbers, each $\exists (A_i \wedge \neg A_i)$ is an abnormality that does not affect standard numbers, and $n \geq 1$ in view of Theorem 5. Every **APA**-model verifies at least one disjunct of every such disjunction—it verifies a $\varphi \in \Phi(\Gamma_1)$—and if the disjunction is minimal, each disjunct is verified by an **APA**-model. We obviously have to consider only disjunctions of the form (33) that are minimal and in which the disjuncts are maximally specific.

We need a survey of the abnormalities that should be taken into account. One would expect to obtain this survey from the models of Γ_1. I have shown that Γ_1 has no models of type 1 or of types 3–4. I have not shown, however, that the four types of models exhaust the models of Γ_1. Also, it is not clear in which way a full set of such types may be obtained or along which lines one might argue that a given set is complete.

There is a different way to classify all models of Γ_1. If a model verifies an inconsistency, it either classifies two intuitively different terms as identical or it classifies two intuitively identical terms as different. The distinction is easily recognized in formal terms. A model verifies

$$\exists x \neg\, x = x \quad (34)$$

iff it classifies some numbers that are intuitively identical as different. A model verifies

$$\exists x \exists y\, x + y' = x, \quad (35)$$

iff it classifies some intuitively different numbers as identical—it identifies the smaller x with the larger $x + y'$. A model that verifies an inconsistency, obviously verifies either (34) or (35).

---

[38] For example a type 2 model of Γ_1 with cardinality larger than $m + 1$ and that does not verify superfluous abnormalities. That no disjunction of such abnormalities is a member of $Cn_{\mathbf{CLuNs}}(\Gamma_2)$ is shown by the same reasoning (invoking a type 2 model of Γ_2 or a type 4 model of Γ_2).

While these considerations lead to a sensible conclusion, there is a faster way to reach that conclusion. The structure of $\mathcal{L}_a$ and the structure of **CLuNs**-models ensure that a model of Γ_1 is inconsistent iff it classifies some numbers as both identical and different. So it classifies some number as different from itself by (16). It follows by P8, (22), and (23) that there is a unique number ℓ that is its own successor and is different from itself.[39] It also follows by (9)–(12) and (24) that all other numbers—those consistently different from ℓ—are smaller than ℓ and linearly ordered by $<$ and that ℓ is the successor of the largest one of them.[40] In view of this and obvious properties of the smallest number, it follows that all models of Γ_1 are type 2 models (with the trivial model M_3 as an extreme case).

We are still considering Case 3, supposing that **PA** is inconsistent. We have seen that, as an effect of P8, all models of Γ_1 are of type 2. It is instructive to check that every subset of $\mathcal{W}_a$ has a type 2 model. However, if $\exists x(x' = 0 \wedge \neg x' = 0)$ would be an **APA**-theorem, then **APA** would be trivial—it would only have type 2 models with empty tail, viz. models isomorphic to M_3. Let us now return for a moment to (33). If P8 were not an **APA**-axiom,

$$\exists x \exists y (x + y' = x \wedge \neg x + y' = x) \vee (0'' = 0'' \wedge \neg 0'' = 0'') \tag{36}$$

might be an **APA**-theorem that is a minimal disjunction of abnormalities. Precisely this is excluded by P8. Every model of Γ_1 that verifies $0'' = 0'' \wedge \neg 0'' = 0''$ also verifies $0''' = 0'' \wedge \neg 0''' = 0''$[41] and hence also verifies $\exists x \exists y (x + y' = x \wedge \neg x + y' = x)$. In general, as Γ_1 has only type 2 models, all models of Γ_1 verify the abnormalities deriving from the fact that an object is different from itself (which is expressed in terms of v) and is its own successor. As a result, in every **APA**-theorem of the form (33), (i) $n = 1$ because all **APA**-models verify the same abnormalities not affecting standard numbers and (ii) $m = 0$ because every model that verifies an abnormality affecting standard numbers also verifies $\exists (A_1 \wedge \neg A_1)$—keep property 3.6 in mind.

Fact 9 *If a disjunction of abnormalities is an **APA**-theorem, then one of its disjuncts does not affect standard numbers and is itself an **APA**-theorem.*

---

[39] As we have seen, ℓ is identical to itself, smaller than itself, larger than itself, and different from itself.

[40] Some of the theorems listed above, for example (26)-(30), express direct consequences of this situation.

[41] A model that verifies $0'' = 0'' \wedge \neg 0'' = 0''$, verifies $0''' = 0''' \wedge \neg 0''' = 0'''$ (by **CLuNs** and P2) and hence also $0''' = 0''$ by P8.

At this point, we are able to address the question what the **APA**-models look like. For a start, they are all type 2 models. So, in view of P8, an **APA**-model verifies $\neg t = t$ iff t names the number that is larger than all numbers (including itself). This object is its own successor: the model verifies $t' = t$. Consider a model M of Γ_1 that has cardinality 18. This model verifies two sorts of abnormalities: (i) those that affect standard numbers and (ii) other (quantified) abnormalities. As far as (i) is concerned, M verifies $\exists (t_1 = t_2 \wedge \neg t_1 = t_2)$ whenever each of $t_1, t_2 \in \mathcal{T}_a$ is intuitively identical to one of the numerals in $\{17, 18, \ldots\}$. The other abnormalities, from (ii), that are verified by M are **CLuNs**-derivable from the abnormalities affecting standard numbers. Examples are $\exists x(x = x \wedge \neg x = x)$, $\exists x \exists y(x = x + y' \wedge \neg x = x + y')$, $\exists x(x = x + 0^{(n)} \wedge \neg x = x + 0^{(n)})$ for all $n \in \mathbb{N}$, etc. As explained in the previous section (in terms of the windlass), if $M^\dagger$ is a model of Γ_1 and has cardinality 19, then it is less abnormal than M, viz. $Ab(M^\dagger) \subset Ab(M)$, because $M^\dagger$ verifies the same abnormalities from (ii), but from (i) verifies only the abnormalities in which t_1 as well as t_2 are intuitively identical to one of the numerals in $\{18, 19, \ldots\}$. Less and less abnormal models of Γ_1 are obtained by raising the cardinality of the model (pulling more knots from the windlass). There is a limit to this, obtained when (after a real hard pull to the rope) a non-standard number is its own successor. Such models of Γ_1 verify no abnormality that affects standard numbers; they only verify abnormalities that are verified by all models of Γ_1.[42] So these are the **APA**-models.

Summarizing: if **PA** is inconsistent, all **APA**-models are type 2 models in which a non-standard number is its own successor. An immediate result of this is that all standard numbers behave exactly as **PA** intends them to behave.

Fact 10 *Where $t_1, t_2 \in \mathcal{T}_a^c$ and $A \in \{t_1 = t_2, \neg t_1 = t_2\}$, $\vdash_{\mathbf{APA}} A$ iff $A \in $ CID.*

Let $\mathcal{W}_a^p$ comprise the variable-free members of $\mathcal{W}_a$. Call S a *maximal consistent subset* of $\mathcal{W}_a^p$ iff, for every $A \in \mathcal{W}_a^p$, S contains exactly one out of $\{A, \neg A\}$.

Theorem 6 *$\{A \in \mathcal{W}_a^p \mid \vdash_{\mathbf{APA}} A\}$ is a maximal consistent subset of $\mathcal{W}_a^p$.*

---

[42] (i) Most premise sets entail a set of disjunctions of abnormalities, but Γ_1 entails a set of abnormalities. So $\Phi(\Gamma_1)$ is a singleton. This exceptional situation is the effect of P8. (ii) Infinite type 2 models verify $\exists x(x^{(n)} = x^{(n)} \wedge \neg x^{(n)} = x^{(n)})$ for every $n \in \mathbb{N}$, but so do finite type 2 models.

Proof. It is easily seen that CID comprises the **APA**-theorems that are members of $\mathcal{W}_a^a$ or negations of those members. So, in view of Fact 5, $Cn_{\mathbf{CL}}(\text{CID}) \cap \mathcal{W}_a^p$ is a maximal consistent subset of $\mathcal{W}_a^p$.

Consider an $A \in Cn_{\mathbf{CL}}(\text{CID}) \cap \mathcal{W}_a^p$. In view of (2.12), CID $\vdash_{\mathbf{CLuNs}}$ $A \vee (B_1 \wedge \neg B_1) \vee \ldots \vee (B_n \wedge \neg B_n)$ in which, for each $B_i \wedge \neg B_i$, either $B_i \in$ CID or $\neg B_i \in$ CID but not both. So by Theorem 5 and Fact 9, no $B_i \wedge \neg B_i$ is a disjunct of a minimal *Dab*-formula that is an **APA**-theorem. It follows that all **APA**-models falsify $(B_1 \wedge \neg B_1) \vee \ldots \vee (B_n \wedge \neg B_n)$ and verify A, whence $A \in \{A \in \mathcal{W}_a^p \mid \vdash_{\mathbf{APA}} A\}$. So $Cn_{\mathbf{CL}}(\text{CID}) \cap \mathcal{W}_a^p \subseteq \{A \in \mathcal{W}_a^p \mid \vdash_{\mathbf{APA}} A\}$.

Suppose that there is an A such that $A \in \{A \in \mathcal{W}_a^p \mid \vdash_{\mathbf{APA}} A\}$ and $A \notin Cn_{\mathbf{CL}}(\text{CID}) \cap \mathcal{W}_a^p$. As $Cn_{\mathbf{CL}}(\text{CID}) \cap \mathcal{W}_a^p$ is a maximal consistent subset of $\mathcal{W}_a^p$, $\{A \in \mathcal{W}_a^p \mid \vdash_{\mathbf{APA}} A\}$ is inconsistent. This is impossible in view of Theorem 5. So $\{A \in \mathcal{W}_a^p \mid \vdash_{\mathbf{APA}} A\} \subseteq Cn_{\mathbf{CL}}(\text{CID}) \cap \mathcal{W}_a^p$. ∎

6 Comments and More Results

This section contains a set of comments that are meant to clarify the results and their significance. Here and there a new result surfaces.

(A) Given the subject, it is obviously important to state that I have relied on all classical means in the metalanguage. To do so is not incoherent. If a theory T that is inconsistent, because both $\vdash_T A$ and $\vdash_T \neg A$ for some A, it is very well possible that the metatheory if T may be phrased within a consistent framework, which for example requires that there is no A such that $\vdash_T A$ and $\nvdash_T A$.

There is, however, a different danger. For example, I have supposed that there are non-standard type 2 models and that Γ_1 has such models because it has finite type 2 models. However, non-standard models impose rather heavy requirements on sets and orderings, see again Boolos, Burgess, and Jeffrey (2002). If **PA** is inconsistent, then presumably most mathematical theories will be inconsistent. I have not shown that, in that case, a coherent set theory results, nor what it looks like, nor that it allows one to define non-standard type 2 models.[43] It seems to me that my results are coherent with the classical framework that I presupposed.[44] To prove this claim correct may involve some pitfalls, but take into account that inconsistency-adaptive logics interpret premise sets as consistently as possible, isolate inconsistencies, and behave like

---

[43] Introducing a paraconsistent logic in the metalanguage is easy enough, but does not help much. One needs at least a rich set theory.

[44] This framework is more confined than one might expect. For example, I nowhere supposed that the standard model or non-standard **PA**-models exist—if **PA** is inconsistent, it has no **CL**-models at all and I did not suppose that it has some.

CL for all reasoning that does not involve one of those inconsistencies.

All this does not make the result dubious. Rather, it places them in the right perspective: they are provable by relying on the commonly accepted classical means.

(B) The axiom P8 has a rather drastic effect: if **APA** is inconsistent, it has only type 2 models. If the axiom were replaced by $\exists x \exists y\, x + y' = x \supset \exists x\, x' = x$ all type 4 models are eliminated except for those that are also type 2 models, all type 3 models are eliminated except for the trivial model M_3,[45] but inconsistent type 1 models are retained. Some other models are also retained, for example models with the following successor graph:

$$\mathfrak{f} \to \ldots \to \mathfrak{a} \to \mathfrak{b} \circlearrowleft \qquad \mathfrak{c} \circlearrowleft$$

$\mathfrak{a}$ and $\mathfrak{b}$ non-standard numbers
$\mathfrak{x} + \mathfrak{c} = \mathfrak{c} + \mathfrak{x} = \mathfrak{c}$ for all $\mathfrak{x}$
$\mathfrak{f} \times \mathfrak{c} = \mathfrak{c} \times \mathfrak{f} = \mathfrak{f}$
$\mathfrak{x} \times \mathfrak{c} = \mathfrak{c} \times \mathfrak{x} = \mathfrak{c}$ for all $\mathfrak{x} \neq \mathfrak{f}$

which is just like an infinite type 2 model of Γ_1, except for the 'free floating' $\mathfrak{c}$. Actually, the number of free floating objects may be multiplied: extend the domain with $\mathfrak{d}$ and copy the listed equations on $\mathfrak{c}$, systematically replacing $\mathfrak{c}$ by $\mathfrak{d}$. This may be repeated indefinitely.

If axiom P8 were replaced by $\exists x \forall y (\exists z\, y + z' = y \supset y = x)$, type 4 models are eliminated except for those that are also type 2 models, but the other types of models of the axioms would be retained. The reason to choose P8 is its effect that disjunctions of quantified abnormalities of the form (33) have only one disjunct.

The use of "choose" in the previous paragraph should be understood correctly. Γ_2 has type 2 models. So eliminating all models of Γ_2 that are not of type 2 does not result in triviality. For other choices, it might be quite difficult to prove that the result is non-trivial. The choice for a type 4 model of which the cycle has period 2 would result in triviality if $\exists x\, x' = x$ is derivable.

(C) We have seen that, in the context of **APA**, triviality can be expressed by $\forall x \forall y\, y = x$ (or by the equivalent $\exists x \forall y\, y = x$). So, in the presence of material implication, classical negation can be defined:

$$\sim A =_{df} A \supset \forall x \forall y\, y = x\,.$$

Note that $\vdash_{\mathbf{APA}} \sim A \supset \neg A$.

An effect of the feature is that $\vdash_{\mathbf{APA}} \sim A$ whenever A is falsified by all **APA**-models. Compare: $\nvdash_{\mathbf{APA}} 0' = 0$ holds as soon as some

---

[45] As mentioned before, it does not make any difference whether M_3 is eliminated or not.

APA-models falsify $0' = 0$, but $\vdash_{\textbf{APA}} \sim 0' = 0$ holds only if all **APA**-models falsify $0' = 0$.[46] Another effect, that will prove useful, is that the consistent behaviour of a formula may be expressed—see for example Batens (1980). Thus if $\sim A$ [$\sim \neg A$] is an **APA**-theorem, then **APA** would be trivial if A [$\neg A$] were also an **APA**-theorem. But we know that **APA** is non-trivial. So $\sim A$ states that A is *consistently false* and $\sim \neg A$ states that A is *consistently true*.

(D) The presence of classical negation throws some light on the meaning of the axioms. E.g., P2 is **CLuNs**-equivalent to $\forall x \forall y (\sim \neg x' = y' \supset \sim \neg x = y)$, which says: if the successors of x and y are *consistently* identical, then so are x and y. In a sense this corresponds to the meaning of P2a in the context of **PA**. P2a to the contrary is **CLuNs**-equivalent to $\forall x \forall y (\sim x = y \supset \sim x' = y')$, which says: if x and y are *consistently* different, then so are their successors. If **PA** is inconsistent, this is not an **APA**-theorem and it is actually false in all models of Γ_1.

(E) Theorems of **PA** may be seen as coming in two kinds. The *desired* ones are those that are theorems in case **PA** is consistent. If **PA** is inconsistent, all other members of $\mathcal{W}_a$ are also **PA**-theorems and will be called *undesired*.

It is not clear whether this distinction can be made sharp. Even if it cannot, what follows suggests that, if **PA** is inconsistent, then there is an interesting and sharp distinction that resembles the one between desired and undesired **PA**-theorems.

Let us suppose that **PA** is inconsistent—the supposition will be kept until we reach (F). The **APA**-models are then non-standard type 2 models. With the exception of the number that is its own successor, numbers behave in these models as they are supposed to behave in classical non-standard models. The presence of classical negation enables one to phrase sentences that make claims about the 'consistent numbers' that occur in the **APA**-models. I shall call these sentences restricted **APA**-theorems. They are obtained from **PA**-theorems by a double transformation. On the one hand, a specific restriction is introduced, stating that certain terms do not refer to 'the inconsistent number' or 'the number that is its own successor'. On the other hand, consistency is explicitly affirmed and in this way the original meaning is restored within the paraconsistent context. Let us first consider some examples.

Note that $\forall x (x' = x \supset (x < x \wedge \neg x < x))$. Let $!t$ abbreviate

---

[46] If $\vdash_{\textbf{APA}} 0' = 0$ holds together with $\nvdash_{\textbf{APA}} 0' = 0$, the metatheory is inconsistent. If $\vdash_{\textbf{APA}} 0' = 0$ holds together with $\vdash_{\textbf{APA}} \sim 0' = 0$, **APA** is trivial. The two are at best contextually equivalent.

$\sim t' = t$—the contextually equivalent $\sim\neg t = t$ would also do.

$$\forall x \sim x' = 0 \tag{37}$$
$$\forall x(!x \supset \sim x^{(n)} = x) \quad (n > 0) \tag{38}$$
$$\forall x \forall y(!x \supset (x \leq y \supset \sim y < x)) \tag{39}$$
$$\forall x \forall y(!x \supset \sim\neg x < x + y') \tag{40}$$
$$\forall x \forall y \forall z((!(x+z) \wedge !(y+z)) \supset (x < y \supset \sim\neg x + z < y + z)) \tag{41}$$
$$\forall x \forall y((!x \wedge !y) \supset (\exists z\, x + z' = y \supset \sim\exists z\, y + z = x)) \tag{42}$$
$$\forall x(!x \supset \exists y(\sim\neg x' = y \wedge \sim x = y)) \tag{43}$$
$$\forall x(!x \supset (\sim\neg x = 0 \vee \exists y(\sim\neg y' = x \wedge \sim x = y))) \tag{44}$$
$$\forall x \forall y((!x \wedge !y) \supset (x' = y' \supset \sim\neg x = y)) \tag{45}$$

Note that $\sim\neg(A \supset B)$ is **CLuNs**-equivalent to $A \supset \sim\neg B$ and that $\sim\neg(A \supset \neg B)$ is **CLuNs**-equivalent to $A \supset \sim B$. The defined classical negation may be replaced by the standard negation in front of an implicatum (in view of $\vdash_{\mathbf{APA}} \sim A \supset \neg A$),[47] which nearly always comes to a weakening of the **APA**-theorem. No restriction is required in (37). (45) is a correctly restricted form of P2a. Unlike what was suggested sub (D), there is no need to require that x' and y' are consistently identical.[48] (43) is **CLuNs**-equivalent to $\forall x(!x \supset \exists y(\sim\neg x' = y \wedge \sim x = y))$. Note, in connection with (40) that $\forall x \forall y\, x < x + y'$ is also an **APA**-theorem, but that $\forall x \forall y \sim\neg x < x + y'$ is not.

The examples (37)–(45) are restricted **APA**-theorems that correspond to desired **PA**-theorems. In order to see what happens to undesired **PA**-theorems, we need to know the precise system behind the restrictions. I have no proof that the system I shall describe is fully adequate. Also, several approaches may work equally well. So what follows is one of the possible approaches and the approach was insufficiently studied.

Let A be a **PA**-theorem and let S be the set of all terms that occur in A. The idea is that $x + 0', x, 0, 0' \in S$ if $x + 0'$ occurs in A. The transformation happens stepwise and is applied to quantified formulas that occur in A, starting with the outermost ones. Every quantified formula $\forall \alpha B$ $[\exists \alpha B]$ is replaced by $\forall \alpha(X \supset B)$ $[\exists \alpha(X \wedge B)]$ for a specific X that is the *restriction* imposed upon the subsentence B of A. In

---

[47] Put differently, the replacement is allowed in positive parts of the sentences—see Schütte (1960).

[48] If neither x nor y is the number that is its own successor and the successors of x and y are identical, then these successors are either consistently identical or they are both identical to the number that is its own successor. Whichever obtains, x and y are consistently identical.

each case, X is the conjunction of the formulas $\sim t' = t$ that fulfil the following requirements: (i) $t \in S$, (ii) t occurs in B, (iii) α occurs in t, and (iv) all variables that occur in t are bound by a quantifier in the transformed formula.[49] Whenever $t \in \mathcal{T}_a^c$, the connected $\sim t' = t$ is redundant in view of Theorem 5. After the described transformations, $\sim\neg$ is inserted before every literal that does not belong to a restriction.

As (37)–(45) illustrate, the restriction may nearly always be weakened in the case of desired **PA**-theorems. This is so either because no restriction needs to be imposed on some term or because one restriction entails the other, or because a weaker restriction will do, for example $!x$ instead of $!x'$.

Let us consider some undesired **PA**-theorems (still supposing that **PA** is inconsistent). All conjuncts of the restrictions are retained for obvious reasons.

$$\sim\neg\, 0' = 0 \tag{46}$$

$$\exists x(\sim x' = x \wedge \sim x = x) \tag{47}$$

$$\exists x(\sim x' = x \wedge \sim\neg\, x' = x) \tag{48}$$

$$\forall x((!x \wedge !0 \wedge !0' \wedge !0'' \wedge !x + 0'') \supset \sim\neg\, x + 0'' = x) \tag{49}$$

$$\forall x(\sim x' = x \supset \exists y((\sim y' = y \wedge \sim(x+y)' = x+y) \wedge x+y = y)) \tag{50}$$

$$\exists y(\sim y' = y \wedge \forall x((\sim x' = x \wedge \sim(x+y)' = x+y) \supset x+y = y)) \tag{51}$$

Neither of these restricted sentences is an **APA**-theorem. (46) is the transformation of the undesired $0' = 0$. Varying the transformation method may result in a slightly different transformation without changing the outcome; for example $(!0 \wedge !0') \supset \sim\neg\, 0' = 0$ is equivalent to (46). (47) is the transformation of $\exists x \neg x = x$, which incidentally is an **APA**-theorem. The transformation is not an **APA**-theorem. It moreover causes triviality in view of $\vdash_{\mathbf{APA}} \forall x(x' = x \vee x = x)$; note that $\forall x(x' = x \vee \sim\neg\, x = x)$ is also an **APA**-theorem, but that $\forall x(\sim\neg\, x' = x \vee \sim\neg\, x = x)$ is not. Another sentence that is an **APA**-theorem if **PA** is inconsistent, is $\exists x\, x' = x$. Its transformation (48) is not an **APA**-theorem and actually causes triviality. (49) is the transformation of the undesired **PA**-theorem $\forall x\, x + 0'' = x$. (50) and (51) are also transformations of undesired **PA**-theorems, respectively $\forall x \exists y\, x + y = y$ and $\exists y \forall x\, x + y = y$. Incidentally, both are **APA**-theorems, but the transformations are not.

The following example seems to raise doubt about the transformation

---

[49] If $\sim t' = t$ is a conjunct of the restriction X in $\forall \alpha(X \supset B)$ [$\exists \alpha(X \wedge B)$], then $\sim t' = t$ will not be a conjunct (for this t) of the restriction of any subformula of B.

method:
$$\forall x((x'' = x' \wedge \sim x' = x) \supset \exists y(y \times 0'' = x)). \tag{52}$$

This sentence states that the predecessor of 'the inconsistent number' is even. It is **APA**-contingent (always supposing that **PA** is inconsistent) and its transformation is

$$\forall x(X \supset (Y \supset \exists y((!y \wedge !(y \times 0'')) \wedge \sim \neg y \times 0'' = x))), \tag{53}$$

in which X abbreviates $(!x \wedge !x' \wedge !x'')$, Y abbreviates $(x'' = x' \wedge \sim x' = x)$, and constant terms are omitted from the second restriction (in view of the redundance). (53) is an **APA**-theorem. Indeed, the first restriction $(!x \wedge !x' \wedge !x'')$ is (classically) contradicted by $x'' = x' \wedge \sim x' = x$. So (53) is a **CLuNs**-theorem, a variant of *ex falso quodlibet*. The outcome does obviously not indicate a weakness of **APA**. Nor does it cast doubt on the transformation method. Indeed, (52) is a *desired* **PA**-theorem because it is a **CL**-consequence of the desired **PA**-theorem $\forall x \neg x'' = x'$. If the situation still raises suspicion, consider the following. The transformation method aims at turning desired **PA**-theorems into statements about the consistent numbers of type 2 models. In sentences like (52) an antecedent refers to a number that is its own successor or to one that is different from itself. They are **PA**-theorems because either **PA** is trivial or the antecedent is false. The transformation of such sentences are **APA**-theorems because an antecedent refers to a number that is its own successor amongst the numbers that are (consistently) not their own successor.

The upshot of the present comment (E) seems to be that, if **PA** is inconsistent, then **APA** distinguishes between desired and undesired **PA**-theorems.

(F) We have seen that, if **PA** is inconsistent, only some **PA**-theorems have a counterpart in **APA**, and actually a counterpart that often involves a restriction. Some readers may find this odd in view of claims, made by Van Bendegem (1993, 1994) where a finite inconsistent arithmetic is defined in terms of type 2 models, that this arithmetic contains all **PA**-theorems. The matter becomes understandable if one realizes (i) that Van Bendegem supposes that **PA**-is consistent and (ii) that, in his papers, the underlying logic is **LP** from Priest (1979, 2006), which does not contain a detachable implication. Thus axiom P2a comes, in my terminology, to $\forall x \forall y (\neg x' = y' \vee x = y)$. This is obviously verified by all type 2 models because if a names the number that is its own successor and b names its predecessor, then $v_M(\neg a' = b') = 1$ for every type 2 model M. Those models, however, falsify $\forall x \forall y(x' = y' \supset x = y)$ because, where A and b are as above, $v_M(a' = b') = 1$ whereas $v_M(a = b) = 0$.

(G) Gödel's First Incompleteness Theorem is directly connected with the fact that **PA** was seen as an attempt to axiomatize the set of sentences verified by a single model, the so-called standard model of arithmetic. The situation is different for **APA** if **PA** is inconsistent. Indeed, if that is so, there are many different **APA**-models. The predecessor of ℓ can be identified: the x such that $x'' = x' \wedge \sim x' = x$. In some **APA**-models, this x is even, in other **APA**-models it is odd but divisible by 3, and so on *ad infinitum*.

These differences may obviously be multiplied. One may also identify the second predecessor of ℓ: the x such that $x''' = x'' \wedge \sim x'' = x'$. Dependent on the factors of the predecessor of ℓ, the second predecessor of ℓ may have or not have certain factors. And so on.

This may be repeated for the third predecessor of ℓ, etc. The set of factors of each of those numbers will contain non-standard numbers. Each of these will have a first predecessor, which will have factors, and so on *ad nauseam*, which only sounds better than the German equivalent.

(H) It is sometimes difficult to figure out which models are retained. Consider the example M_5:

$\mathfrak{f} \to \mathfrak{a} \to \mathfrak{b} \to \ldots \quad \mathfrak{n} \circlearrowright$

only $\mathfrak{n}$ non-standard
$\mathfrak{x} + \mathfrak{n} = \mathfrak{n} + \mathfrak{x} = \mathfrak{n}$ for all $\mathfrak{x}$
$\mathfrak{f} \times \mathfrak{n} = \mathfrak{n} \times \mathfrak{f} = \mathfrak{f}$
$\mathfrak{x} \times \mathfrak{n} = \mathfrak{n} \times \mathfrak{x} = \mathfrak{n}$ for all $\mathfrak{x} \neq \mathfrak{f}$

The non-standard number $\mathfrak{n}$ comes after all standard numbers, as always. However, it is the only non-standard number and so is its own predecessor as well as its own successor. Is M_5 an **APA**-model? Actually, it is not, as the following **CLuNs**-proof shows.

1	$\forall x(x'' = x' \supset x' = x)$	Hypothesis
2	$\forall x((x' = x \supset 0' = 0) \supset (x'' = x' \supset 0' = 0))$	1; $A \supset B/(B \supset C) \supset (A \supset C)$
3	$0' = 0 \supset 0' = 0$	**CLuNs**-theorem
4	$\forall x(x' = x \supset 0' = 0)$	2, 3; P7
5	$\exists x \, x' = x$	Hypothesis
6	$0' = 0$	4, 5; $\forall x(A(x) \supset B), \exists x A(x) / B$
7	$\forall x(x + 0' = x + 0)$	6; $a = b / \forall x(x + a = x + b)$
8	$\forall x(x' = x)$	7; by P3 and P4
9	$0 = 0$	**CLuNs**-theorem
10	$\forall x(x = 0 \supset x' = 0)$	8; $f(x) = x / A(x) \supset A(f(x))$
11	$\forall x \, x = 0$	9, 10; P7
12	$\forall x \forall y \, x = y$	11; $x = a, y = a / x = y$
13	$\exists x \, x' = x \supset \forall x \forall y \, x = y$	5, 12; Conditional Proof
14	$\sim \exists x \, x' = x$	13; Def.$\sim$
15	$\forall x(x'' = x' \supset x' = x) \supset \sim \exists x \, x' = x$	1, 14; Conditional Proof
16	$\exists x \, x' = x \supset \sim \forall x(x'' = x' \supset x' = x)$	15; Contraposition

17 $\exists x\, x' = x \supset \exists x(x'' = x' \wedge \sim x' = x)$ 16; **CLuNs**-equivalences

The conclusion is a **CLuNs**-consequence of $\{P3, P4, P7\}$ and hence of Γ_1. It states that, if a number is its own successor, then there is a number that is (consistently) not its own successor, but the successor of which is (inconsistently) its own successor. So the number that is its own successor is classically different from its predecessor.[50] As M_5 falsifies this, it is not a model of Γ_1 and hence not an **APA**-model.

(I) The obtained results have their price. **APA** is non-trivial and is either identical to **PA** or apparently retains the desired **PA**-theorems in restricted form. The price paid is that the semi-recursive **PA** is replaced by a theory that may be Π^1_1-complex, which is the maximal complexity of **CLuNs**^{m}-consequence sets of recursive premise sets—see Odintsov and Speranski (2012, 2013); Verdée (2009).[51] The complexity results from the dynamic proofs and more precisely from the marking definition for Minimal Abnormality. Actually, in view of Theorem 5 and Fact 9 the computational complexity of **APA** is maximally Σ^0_3.

If **PA** turns out to be inconsistent, the complexity of **APA** is a small price to pay for obtaining a non-trivial theory that is closely related to **PA**. There is, however, a different inconvenience. If, in order to 'play it safe', we have to replace **PA** by **APA**, does this mean that doing arithmetic becomes a more difficult matter? By no means.

There is a close relation between **PA**-proofs and **APA**-proofs. If one deletes the condition from an **APA**-proof,[52] the result is a **PA**-proof. Conversely, a **PA**-proof is turned into an **APA**-proof by adding the conditions.[53] Adding these conditions is a purely algorithmic matter. The fact that **PA** is defined in terms of Γ_0 while **APA** is defined in terms of Γ_1 makes the algorithm only slightly more complicated than if both would share the same axioms.

All this has a dramatic effect. As long as no inconsistency is derived

---

[50] The defined $\sim$ was introduced to save horizontal space. It is instructive to write out the proof without that symbol.

[51] Verdée (2009) reacts on a mistaken proof of a mistaken theorem in Horsten and Welch (2007) and highlights the strengths of the fact that so huge a complexity is captured by a predicative logic. The philosophical contentions and presentation of Horsten and Welch (2007) are criticized in Batens et al. (2009). Odintsov and Speranski (2012, 2013) do not refer to philosophical disagreements, but delineate conditions under which the complexity is reduced.

[52] The marks are not part of the proofs, even if they are often displayed for pedagogical reasons or for reasons of clarity. Whether a line is marked or unmarked is determined by the marking definition, not by any action taken by the author of the proof.

[53] In general, the relation between the two kinds of proofs holds between the dynamic proofs of the adaptive logic and the static proofs of its upper limit logic.

from **PA**, we can simply go on deriving **PA**-theorems. No line has to be marked anyway; whatever is derived can (defeasibly) be considered as finally derived in view of the present insights. If the resulting set turns out inconsistent, we write a computer program and let a computer turn all interesting **PA**-proofs into **APA**-proofs. Figuring out which lines are marked is also an algorithmic matter. So that can also be left to the computer program.

In **CLuNs**^{m}-proofs, final consequences depend on the disjunctions of abnormalities that are **CLuNs**-derivable from the premise set (here axioms Γ_1). As **CLuNs**-consequence sets are in general only semi-recursive, there is no way to know at any point whether all derivable disjunctions of abnormalities have been derived. For **APA**, however, the situation is different. We know beforehand which inconsistencies will be **APA**-theorems if **PA** is consistent: $\exists x(x' = x \wedge \neg x' = x)$ and all abnormalities that are **APA**-derivable from this. The set of these abnormalities comprises the abnormalities verified by finite type 2 models, for example M_2, minus the abnormalities that affect standard numbers. All abnormalities have the form $Q(t_1 = t_2 \wedge \neg t_1 = t_2)$ in which Q is a sequence of quantifiers. The only members of Ω that are **CLuNs**-derivable from $Q(t_1 = t_2 \wedge \neg t_1 = t_2)$ are obtained from it by Universal Instantiation. The abnormality $Q(t_1 = t_2 \wedge \neg t_1 = t_2)$ affects standard numbers iff one of these instantiations turns t_1 or t_2 into a member of $\mathcal{T}_a^c$. To find out whether the condition is fulfilled, it is sufficient to instantiate every universally quantified variable with a different member of a set of dummy variables for numerals, for example $\{0^{(n_1)}, 0^{(n_2)}, \ldots\}$.

Fact 11 *If* **PA** *is inconsistent, there is exactly one* $\varphi \in \Phi(\Gamma_1)$ *and* $\varphi = \Omega \cap Cn_{\mathbf{CLuNs}}(\Gamma_1)$ *is decidable.*

So, in view of Theorem 5 and Fact 9, the following two facts obtain:

Fact 12 *If* **PA** *is inconsistent, a line of an* **APA**-*proof with condition* Δ *is marked iff* $\Delta \cap \varphi \neq \emptyset$ *(with* $\{\varphi\} = \Phi(\Gamma_1)$*).*

Fact 13 *If* **PA** *is inconsistent,* **APA** *is semi-recursive.*

If **PA** is consistent, **APA** is also semi-recursive. As we may forever be unable to settle the question whether **PA** is consistent, we do not know which of two semi-recursive sets is the set of **APA**-theorems. Consider the following two sets, with φ such that $\{\varphi\} = \Phi(\Gamma_1)$ if **PA** is inconsistent:

$\Lambda_1 = \{A \mid \text{for some } \Delta,\ A \vee Dab(\Delta) \in Cn_{\mathbf{CLuNs}}(\Gamma_1) \text{ and } \varphi \cap \Delta = \emptyset\}$
$\Lambda_2 = Cn_{\mathbf{CL}}(\Gamma_1) - \Lambda_1$

Recall that $Cn_{\mathbf{CL}}(\Gamma_1) = \{A \mid \text{for some } \Delta,\ A \vee Dab(\Delta) \in Cn_{\mathbf{CLuNs}}(\Gamma_1)\}$; obviously $\Delta = \emptyset$ is not excluded.

To avoid misunderstanding, I add a comment. If **PA** is consistent, the **APA**-consequence set is $\Lambda_1 \cup \Lambda_2$ because no abnormality will ever be derived whence no line will ever be marked. If **PA** is inconsistent, the **APA**-consequence set is Λ_1 and $\Lambda_1 \cup \Lambda_2$ is trivial. Both sets are semi-recursive. The optimists may start a Turing machine to enumerate $\Lambda_1 \cup \Lambda_2$, the pessimists one to calculate Λ_1 and collaborating will speed up things. If either set turns out to contain an inconsistency, the first Turing machine is switched off.

The conclusion is that, (i) if **PA** is consistent, all and only its theorems are **APA**-theorems and (ii) if **PA** is inconsistent, all its desired theorems seem to correspond to possibly restricted **APA**-theorems. Moreover, for any given **PA**-theorem one may decide right now whether it will have to be restricted if **PA** is inconsistent and which restriction will have to be introduced.

7 In Conclusion

The presented results deserve further study. One of the reasons is precisely that the results are so nice and simple that they raise suspicion. Suppose indeed that **PA** is inconsistent. Even then, **APA** is extremely close to **PA**. In the **APA**-models, all natural numbers have their intended properties—in 'most' models even an infinity of infinities of nonstandard numbers have the same properties. Consistent behaviour is amongst the properties. Inconsistency is isolated within a single nonstandard number,[54] a number that belongs to an infinity of infinities of numbers that no one was interested in in the first place. The reader may object that part of this result is forced on **APA** by axiom P8. This is not an objection, however. What is so remarkable is that P8 does not cause triviality, viz. that all possible inconsistencies can be forced into $\exists x\, x' = x$.

Even syntactically the **PA**-theorems are retained, provided one adds the restriction, for each term that occurs in the theorem, that it is not its own successor. Often no restriction is required. Remarkably enough, this holds for all theorems in which no variables occur and in generalizations obtained from these, for example $\forall x \neg 17 + x' = 17$.

The restriction can be expressed within the language of arithmetic. Moreover, whenever A is an **APA**-theorem and its negation is not, this

---

[54] Last warning: where x is the number that is its own successor and y is such that $y \times 0'' = x$ there is no inconsistency in y or in 0, $0'$ or $0''$, but only in $y \times 0''$, which actually is x.

can be expressed within the language of arithmetic and this expression is an **APA**-theorem, viz. $\sim\neg A$. Note that this gives the consistent **APA**-theorems exactly the same force as **PA**-theorems. For example, the fact that $\sim 0' = 0$ is an **APA**-theorem excludes that $0' = 0$ is an **APA**-theorem, just as much as the fact that $\neg 0' = 0$ is a **PA**-theorem. In both cases, the provability of $0' = 0$ would cause the triviality of the involved theory. As said before, this closeness of **APA** and **PA** underlines the need for further studying the results.

There are many open problems. Tackling these seems interesting and promising in view of the preceding paragraphs.

▷ It seems very important that the reasoning leading to the results is made more transparent. This will reveal which specific presuppositions are involved and which means have been used. Such information will be useful for figuring out where an inconsistency might come from, if one is ever derived from **PA**. Incidentally, one should not forget the means applied within the metatheory of **CLuNs**^{m} in as far as it is invoked for the results on **APA**.

It should be mentioned in this connection that I often proceeded in terms of models to make my and the reader's life simpler. Any reference to infinite models obviously undermines the robustness of the results because infinite models presuppose many of the matters under discussion. So rewriting the paper in proof theoretic terms is a task of the highest priority.

▷ It is not clear what becomes of Gödel's First Incompleteness Theorem. Classical negation is definable in **APA**, the set of **APA**-theorems is one of two semi-recursive sets, Λ_1 and $\Lambda_1 \cup \Lambda_2$, and, for each set, the proof that A belongs to the set is finite and so may be given a finite Gödel number.[55] So a proof that a sentence belongs to Λ_1, respectively $\Lambda_1 \cup \Lambda_2$ is apparently representable in terms of restricted theorems within Λ_1. But then Gödel sentences are definable and they will neither be provable nor classically disprovable. There is no need to add "if **APA** is non-trivial" because we know that it is. Many questions remain, however, for example some connected with the standard paraconsistent negation.

Many similar questions can be raised in connection with Gödel's second Incompleteness Theorem, obviously with "consistent" replaced by "non-trivial". We have seen that **APA** is non-trivial, but the used proofs have to be better checked. It is easy enough to see that the

---

[55]I obviously refer to the proofs described in Fact 12. The presence of infinite proofs would obviously not introduce non-standard Gödel numbers, but would make Gödel numbering impossible.

$Cn_{\mathbf{CLuNs}}(\Gamma_1)$ is non-trivial in view of its finite models. Going from there to the non-triviality of **APA**, I relied on (3.7). The available proof of this refers to infinite models and infinite sets—because it is derived from the much stronger Strong Reassurance (Stopperedness, Smoothness). However, the existence of a more absolute proof in terms of the proof theory seems plausible.

▷ If an inconsistency is derived from **PA**, one should figure out whether Γ_0 may be changed in order to avoid the inconsistency and, if so, in which ways. As all natural numbers and 'nearly all' non-standard numbers behave consistently, it would be odd, although not impossible, that no sensible theory would avoid the inconsistency.

▷ Part of what precedes is presumably not optimal. For example, there may be better alternatives for transforming **PA**-theorems into restricted **APA**-theorems. Actually, even if the proposed transformation is optimal, it deserves a careful metatheoretic study. Indeed, what is said *sub* (E) in Section 6 suggests that, even if **PA** is inconsistent, a theory close to **PA** can be delineated.

▷ For a while I hoped to find a definition of desired **PA**-theorems in terms of the prospective dynamics described in Batens (2012), which itself grew from an insight from Batens and Provijn (2001). Proof search is goal-directed and so are cleaned-up proofs. In Batens and Provijn (2001) a proof procedure was devised for propositional **CL** in an attempt to do justice to the specific goal-directedness of proofs—the predicative version is being prepared by Peter Verdée and Dagmar Provijn. It turned out (at the propositional level) that a sensible and natural proof search procedure is readily available, but that one has to introduce a special device to obtain the consequences of *Ex Falso Quodlibet*, also known as *Explosion*. The feature was by no means intended, but turned out unavoidable—see Batens (2012) for the fascinating procedure without Explosion. The main lesson of this is that the intrinsic entanglement of Explosion in **CL** is a property of the Tarski consequence relation, more specifically of transitivity, but is not a property of the proof search procedure in **CL**. The matter is similar for adaptive logics; see Batens (2005) and Verdée (2013) for a generalizable propositional case.

So it might be expected that the procedure without explosion would rule out most undesired **PA**-theorems. Not all of them, for it is bound to lead to an inconsistent set of theorems. Although the predicative version is not fully settled at this moment, it is extremely likely that, if **PA** is inconsistent, then all members of $\mathcal{W}_a$ are derivable by the procedure without explosion. The argument for this claim derives from a finding of Peter Verdée's. Consider the following instances of P7.

$$(0' = 0 \supset A) \supset (\forall x(x' = x \supset A) \supset (x'' = x' \supset A)) \supset \forall x(x' = x \supset A))$$
$$(\neg 0 = 0 \supset A) \supset (\forall x((\neg x = x \supset A) \supset (\neg x' = x' \supset A)) \supset \forall x(\neg x = x \supset A))$$

It seems unavoidable that, for both sentences, the first antecedent will be obtained from P1 (respectively by pure logic) and the second antecedent is obtained from P2. As a consequence, any sentence A will be obtained if $\exists x \, x' = x$ or $\exists x \neg x = x$ can be obtained. The point is, of course, that all this can apparently be obtained by means of the procedure that does not validate explosion.

Spelling out the predicative version of the procedure is certainly urgent. If this version equates inconsistency with triviality for **PA**, it may still be possible to avoid this equation by imposing a restriction upon the procedure. The restriction might for example pertain to inconsistent sets of obtained sentences and refer to the (partial) order in which sentences may be obtained from the axioms.

▷ Is it possible to detect, in terms of (the predicative extension of) the procedure from Batens (2012), which axioms may allow one to derive either $\exists x \neg x = x$ or $\exists x \, x' = x$? This information might be rather easy to obtain and seems to rely on simple and finitistic operations. The information will provide insights in the way in which the inconsistencies are interleaved within **APA**. Perhaps they may even enable one to rephrase the axioms in such a way that the inconsistencies are not derivable.

▷ Most of the results have presumably a bearing on the consistency of arithmetic. This deserves careful checking in several respects. The relevant information involves most of what was said in the previous paragraphs. Further relevant questions concern the representation of parts of the metatheory of **APA** within **APA** and perhaps even their representation within **PA**.

▷ Our insight in **APA** would be furthered iff we had a precise description of the $\varphi \in \Phi(\Gamma_1)$ and of the **CLuNs**-entailment relations between all its members. This task is simple and standard.

▷ The following problem is related to the previous and can be solved by rather standard methods. Part of the complexity, for us, of **APA** derives from the fact that we do not know whether **PA** is consistent. However, **APA** has an extension that is obviously semi-recursive. It is obtained by adding the axiom $\exists x \, x' = x$ to Γ_1. This will lead to different sets of consequences of the so extended **APA**: (i) **CLuNs**-consequences of the axioms—these are derivable on the condition $\emptyset$ in the dynamic proofs, (ii) **PA**-theorems A for which $\sim\neg A$ is an **APA**-theorem, (iii) **PA**-theorems that are **APA**-theorems, (iv) **PA**-theorems the transformation of which are restricted **APA**-theorems—an interesting and easy task concerns the reduction of the restriction, and so on.

As the extended **PA** is semi-recursive, solving this problem should not be very difficult. One should basically look up **PA**-proofs, transform them to **APA**-proofs, mark the lines of these proofs, and add new lines with restricted forms of the **PA**-theorems that occur at marked lines, trying to make the restriction as weak as possible. All this is algorithmic. While carrying the task out, one may experiment, for example with different restrictions.

▷ If A contains only proper terms that refer to numbers smaller than the one that is its own successor, then apparently $A \land \neg A$ cannot be an **APA**-theorem. This suggests that a consistent theory about those numbers may be articulated. Several approaches are possible and each will involve some sophistication.

Among the neater approaches might be one in which the axioms are restricted in terms of a predicate P. P1 might become $\forall x(Px' \supset \neg x' = 0)$, P2 might become $\forall x \forall y((Px' \land Py') \supset (x' = y' \supset x = y))$, and P4 might become $\forall x \forall y(P(x+y)' \supset x + y' = (x+y)')$. Such formulas might be 'interpreted' within the inconsistent extension of **APA**, introduced some paragraphs ago. Pt might be translated into $\sim t' = t$. The new theory should be consistent and the extension of **APA** might be considered as defining an adequacy condition.

This ends the partial list of open problems. Some are not difficult to solve. I hope this attracts some help and I will be grateful for being kept informed.

Allow me to add a theoretical comment. To the best of my knowledge, articulating theories with a defeasible underlying logic is a Ghent idea. I hope that the present paper shows that, in some circumstances, the idea may be rather attractive.

Acknowledgments

Research for this paper was supported by subventions from Ghent University and from the Fund for Scientific Research – Flanders. Parts of this paper were presented in a lecture at CLMPS 14, Nancy, France 2011, and in a lunchtalk at Ghent University, Belgium 2013. I am grateful for comments on a former draft by Rafał Urbaniak, Christian Straßer, and Joke Meheus. I am especially grateful to Peter Verdée and to a referee because they very carefully criticized a former draft.

References

Arieli, O., A. Avron, and A. Zamansky (2011). "Maximal and premaximal paraconsistency in the framework of three-valued semantics". In: *Studia Logica* 97, pp. 31–60.

Avron, A. (1986). "On an implication connective of RM". In: *Notre Dame Journal of Formal Logic* 27, pp. 201–209.

— (1991). "Natural 3-valued logics—Characterization and proof theory". In: *The Journal of Symbolic Logic* 56, pp. 276–294.

Batens, D. (1980). "Paraconsistent extensional propositional logics". In: *Logique et Analyse* 90–91, pp. 195–234.

— (2005). "A procedural criterion for final derivability in inconsistency-adaptive logics". In: *Journal of Applied Logic* 3, pp. 221–250.

— (2007). "A universal logic approach to adaptive logics". In: *Logica Universalis* 1, pp. 221–242.

— (2012). "It might have been Classical Logic". In: *Logique et Analyse* 218, pp. 241–279.

— (201x). *Adaptive Logics and Dynamic Proofs. Mastering the Dynamics of Reasoning*. Forthcoming.

Batens, D. and K. De Clercq (2004). "A rich paraconsistent extension of full positive logic". In: *Logique et Analyse* 185–188, pp. 227–257.

Batens, D. and D. Provijn (2001). "Pushing the search paths in the proofs. A study in proof heuristics". In: *Logique et Analyse* 173–175, pp. 113–134.

Batens, D. et al. (2009). "Yes fellows, most human reasoning is complex". In: *Synthese* 166, pp. 113–131.

Boolos, G. S., J. P. Burgess, and R. J. Jeffrey (2002). *Computability and Logic*. 4th ed. Cambridge University Press.

Boolos, G. S. and R. J. Jeffrey (1989). *Computability and Logic*. 3rd ed. Cambridge University Press.

Carnielli, W. A., J. Marcos, and S. de Amo (2001). "Formal inconsistency and evolutionary databases". In: *Logic and Logical Philosophy* 8, pp. 115–152.

D'Ottaviano, I. M. L. (1982). "Sobre uma Teoria de Modelos Trivalente (in Portuguese)". PhD thesis. State University of Campinas (Brazil).

— (1985a). "The completeness and compactness of a three-valued first-order logic". In: *Proceedings of the 5th Latin American Symposium on Mathematical Logic. Revista Colombiana de Matemáticas*. Vol. 1–2, pp. 77–94.

— (1985b). "The model extension theorems for $\mathbf{J}_3$-theories". In: *Methods in Mathematical Logic: Proceedings of the 6th Latin American*

Symposium on Mathematical Logic. Ed. by C. A. Di Prisco. Vol. 1130. Lecture Notes in Mathematics. Springer-Verlag, pp. 157–173.

D'Ottaviano, I. M. L. (1987). "Definability and quantifier elimination for $\mathbf{J}_3$-theories". In: *Studia Logica* 46.1, pp. 37–54.

D'Ottaviano, I. M. L. and R. L. Epstein (1988). "A paraconsistent many-valued propositional logic: $\mathbf{J}_3$". In: *Reports on Mathematical Logic* 22, pp. 89–103.

Esser, O. (2003). "A strong model of paraconsistent logic". In: *Notre Dame Journal of Formal Logic* 44, pp. 149–156.

Gödel, K. (1931). "Über formal unentscheidbare Sätze der Principia Mathematica und verwandter Systeme I". In: *Monatshefte für Mathematik und Physik* 38, pp. 173–198.

— (1986). *Collected Works*. Ed. by S. Feferman et al. Vol. 1. Oxford: Oxford University Press.

Horsten, L. and P. Welch (2007). "The undecidability of propositional adaptive logic". In: *Synthese* 158, pp. 41–60.

Kleene, S. C. (1952). *Introduction to Metamathematics*. Amsterdam: North-Holland.

Odintsov, S. P. and S. O. Speranski (2012). "On algorithmic properties of propositional inconsistency-adapive logics". In: *Logic and Logical Philosophy* 21, pp. 209–228.

— (2013). "Computability issues for adaptive logics in multi-consequence standard format". To appear; doi:10.1007/s11225-013-9531-2.

Priest, G. (1979). "The logic of paradox". In: *Journal of Philosophical Logic* 8, pp. 219–241.

— (1994). "Is arithmetic consistent?" In: *Mind* 103, pp. 337–349.

— (1997). "Inconsistent models of arithmetic. Part I: Finite models." In: *Journal of Philosophical Logic* 26, pp. 223–235.

— (2000). "Inconsistent models of arithmetic. Part II: The general case." In: *Journal of Symbolic Logic* 65, pp. 1519–1529.

— (2006). *In Contradiction. A Study of the Transconsistent*. 2nd ed. Oxford: Oxford University Press.

Schütte, K. (1960). *Beweistheorie*. Berlin: Springer.

Smirnova, E. D. (2000). "An approach to the justification of semantics of paraconsistent logics". In: *Frontiers of Paraconsistent Logic*. Ed. by D. Batens et al. Baldock: Research Studies Press, pp. 255–262.

Van Bendegem, J. P. (1993). "Strict, yet rich finitism". In: *First International Symposium on Gödel's Theorems*. Ed. by Z. Wolkowski. Singapore: World Scientific, pp. 61–79.

— (1994). "Strict finitism as a viable alternative in the foundations of mathematics". In: *Logique et Analyse* 145, pp. 23–40.

Verdée, P. (2009). "Adaptive logics using the minimal abnormality strategy are Π_1^1-complex". In: *Synthese* 167, pp. 93–104.
— (2013). "A proof procedure for adaptive logics". In: *Logic Journal of the IGPL* 21, pp. 743–766.

CHAPTER 3

How Mathematicians Convince Each Other, or "The Kingdom of Math is Within You"*

Reuben Hersh

1 Introduction

There is a problem about mathematical proof—actual mathematical proof, done by actual mathematicians. Typical samples of mathematicians' proof hardly resemble formal proof.

Formal proof has been well described and analyzed by logicians. Mathematicians' proofs are different. We rarely cite the rules of logic. (Many of us don't know much about logic.) Our "proofs" don't usually start from axioms. If they include rule-governed calculations, these calculations are embedded in non-calculative reasoning—a combination of verbal argument and citation of literature. Nevertheless, these "informal" proofs work—they compel agreement. That is what mathematicians mean by a proof—an argument that compels agreement.

---

*This paper will also be appearing in *Experiencing Mathematics: What do we do, when we do mathematics?*, a collection of existing and new essays on mathematical practice written by this author, to be published by the American Mathematical Society (2014).

Mathematicians and logicians both end by claiming that a certain statement has been "proved". But the steps that reach this claim are very different. Recognizing this discrepancy, Brendan Larvor has asked the following question: *What qualifies the usual informal mathematical arguments as "proof"?* (Larvor 2012).

Before answering Larvor's question, I begin by posing and answering a different question. Not "what qualifies them?", but rather, "What makes informal proof work? How does it compel agreement?"

Mathematicians' arguments are glaringly incomplete from the point of view of formal logic, yet they actually do compel agreement. This "working", this compelling agreement, is stronger in mathematics than in any other human endeavor. Indeed, it is the defining quality of mathematics. "Mathematics is the science that draws necessary conclusions" (Peirce 1879, p. 1).

How does it "work"? How does it compel assent, inconclusive as it is by the standards of formal logic? *Mathematician's proof works for the same reason that ordinary empirical science works.* Different people observing the stars see the same thing, because they are looking at the same thing. Different mathematicians observing a sporadic group or a stochastic process "see" the same thing, because they also are "looking" at the same thing. That is how, starting from established mathematics, we establish a new result, which then becomes part of established mathematics.

This claim is argued below, supported by a classic description of Isaac Newton's thinking, and by half a dozen famous mathematicians, describing their experience of this internal reality.

2 A quote from Hardy

According to Monk (1991), an article by G. H. Hardy exasperated his friend Ludwig Wittgenstein. Hardy's article discussed Russell's logicism, Hilbert's formalism, and Brouwer's intuitionism. Into this dry philosophizing, Hardy threw a strange passage, which some philosophers have condemned as sheer nonsense. That is a pity. Hardy was trying hard to explain what really goes on in mathematicians' proof.

> I have myself always thought of a mathematician as in the first instance an *observer*, a man who gazes at a distant range of mountains and notes down his observations. His object is simply to distinguish clearly and notify to others as many different peaks as he can. There are some peaks which he can distinguish easily, while others are less clear. He sees A

sharply, while of B he can obtain only transitory glimpses. At last he makes out a ridge which leads from A, and following it to its end he discovers that it culminates in B. B is now fixed in his vision, and from this point he can proceed to further discoveries. In other cases perhaps he can distinguish a ridge which vanishes in the distance, and conjectures that it leads to a peak in the clouds or below the horizon. But when he sees a peak he believes that it is there simply because he sees it. If he wishes someone else to see it, he *points* to it, either directly or through the chain of summits which led him to recognize it himself. When his pupil also sees it, the research, the argument, the *proof* is finished.

The analogy is a rough one, but I am sure that it is not altogether misleading. If we were to push it to its extreme we should be led to a rather paradoxical conclusion; that there is, strictly, no such thing as mathematical proof; that we can, in the last analysis, do nothing but *point*; that proofs are what Littlewood and I call *gas*, rhetorical flourishes designed to affect psychology, pictures on the board in the lecture, devices to stimulate the imagination of pupils. This is plainly not the whole truth, but there is a good deal in it. The image gives us a genuine approximation to the processes of mathematical pedagogy on the one hand and of mathematical discovery on the other; it is only the very unsophisticated outsider who imagines that mathematicians make discoveries by turning the handle of some miraculous machine. (Hardy 1929, p. 18)

Of course Hardy knew there was no literal mountain peak that he or his student was seeing! What did he mean? *He meant that he perceived some fact or phenomenon about his internal mathematical world, and he could get another mathematician to perceive it, by guiding him to it through that mathematician's own mental world.*

He and his pupil each had his internal mathematical universe, where he could make observations. And these observations, by introspection, seem definite and indubitable, like our observations of the exterior world by our eyes and ears.

If an ornithologist discovers some unexpected entity in the body of a certain species of swan, his claim is checked by other ornithologists, who have access to samples of that species of swan. They look where he tells them to look, and either they do or don't see what he says is there. Something precisely analogous takes place in mathematicians proof! When Hardy makes a discovery, he explains how other mathematicians can verify his claim, by following a certain sequence of steps, to arrive at "seeing it". And those directions are "the proof"!

Platonism mistakenly locates mathematical entities "out there", in an external unspecified realm of non-human, non-physical reality. But they are right here, in our own individual minds, shared also with many other individual minds. Their reality is both psychological and social, it is mental-cultural.

Mathematical concepts are special, and different from non-mathematical concepts, in having definite properties or attributes, which obtain near unanimous agreement. Some of these properties are immediate, or, as we say, "by definition". Others are not immediate, and require demonstration, which we call "proof".

By "proof", mathematicians mean a procedure which suffices to *establish* some "result" by near-unanimous consent. There are various kinds of proofs.

There are visual proofs, where displaying a diagram or a graph is sufficient. There are directly conjunctive proofs, where the result is "seen" immediately, by combining two or more results from *established mathematics*. And there are proofs of the kind Hardy describes.

Even after Hardy has "seen the peak" (to follow his own metaphor), he may still have a lot of work to do, in order to find and describe a "ridge", along which he can guide his student also to see the peak.

A mathematician trying to prove a theorem has to accomplish two tasks: to find a proof, and to present the proof. Finding the proof means finding the appropriate and relevant mathematical entities or objects, possessing the relevant properties, by means of which to make clear *to himself*, to make himself see, the desired mathematical claim. Then, to convince others, he must somehow arrange the relevant mathematical entities and properties, *put them in order*, so as to talk or write about them successively or repeatedly, and enable his reader or listener also to see what he sees. This orderly presentation is often called "reasoning", or "completing the proof". It's not reasoning about sentences, not syntactic or logical reasoning. It's reasoning about the properties of mathematical objects, properties which may be apparent from one's mental models, or may be cited from the literature.

Sometimes the proof is criticized for an error or a gap. Wiles' first presentation of his proof (more on it below) was found to be incomplete, he worked for another year, with help from Richard Taylor, to complete the proof. Such work of completing or correcting a proof is still based on the properties of mathematical entities, not on formalization or syntactic reasoning. It is done by more complete and careful *mathematical* reasoning.

The mathematician's proof leads the learner to observe and manipulate his/her own mental models, enabling her/him to "see" —to apprehend directly— the claimed attribute or property of the mental model in question. And therefore the learner apprehends *the concept*, the mathematical object, of which that model is *a representative*. We proceed to support these claims, by recent quotes from leading mathematicians, and by a look into the mind of Sir Isaac Newton.[1]

3 What some mathematicians say they are doing

Start with brief but pungent remarks.

> In mathematics you have concrete objects before you and you interact with them, talk with them. And sometimes they answer you. (Heintz 2000, quoted in Harris 2008, p. 974)

> If you can steal ideas, then they are real. Every mathematician knows that ideas can be and often are stolen. (Harris 2008, p. 971)

> When mathematicians refer to "intuition" in the sense I have in mind, it is crucially public ... it can be transmitted from teacher to student, or through a successful lecture, or developed collectively by running a seminar and writing a book on the proceedings. (op.cit., p. 974)

> – Marcel Paul Schutzenberger: 'When Alain Connes and I speak about the reality of mathematical objects, what we mean is that one or another mathematical object is as real to us as the singular nerve ganglion of some tiny creature is to the electrophysiologist who studies it'.
> – Alain Connes: 'Exactly'.
> (Connes, Lichnerowicz, and Schützenberger 2001, p. 41)

> – Schutzenberger: 'I confess I have great difficulty distinguishing my activity from that of an entomologist'.
> – Connes: 'I agree completely'. (op.cit., p. 37)

> Five-dimensional shapes are hard to visualize—but it doesn't mean you can't think about them. Thinking is really the same as seeing. (William Thurston, as quoted in his obituary, *The New York Times*, 22 August 2012)

---

[1]This account of the way "mental models" function can be compared to Jody Azzouni's notion of "inference packages" (Azzouni 2005).

The reader who is not a mathematician may never have heard of William Thurston or Alain Connes. Let's turn to Isaac Newton, described in the famous talk "Newton the Man" by Lord Keynes:

> Until the second phase of his life, he was a wrapt, consecrated solitary, pursuing his studies by intense introspection with a mental endurance perhaps never equaled. [...] I believe that the clue to his mind is to be found in his unusual powers of continuous concentrated introspection. [...] his peculiar gift, especially amongst his contemporaries, was the power of holding continuously in his mind a purely mental problem until he had seen straight through it. I fancy his preeminence is due to his muscles of intuition being the strongest and most enduring with which a man has ever been gifted. Anyone who has ever attempted pure scientific or philosophical thought knows how one can hold a problem momentarily in one's mind and apply all one's powers of concentration to piercing through it, and how it will dissolve and escape and you find that what you are surveying is a blank. I believe that Newton could hold a problem in his mind for hours and days and weeks until it surrendered to him its secret. Then being a supreme mathematical technician he could dress it up, how you will, for purposes of exposition, but it was his intuition which was preeminently extraordinary—"so happy in his conjectures", said De Morgan, "as to seem to know more than he could possibly have any means of proving". The proofs, for what they are worth, were, as I have said, dressed up afterward—they were not the instrument of discovery. (Keynes 1971)

Now turn to Andrew Wiles, famous for proving Fermat's Last Theorem, speaking on a *Nova* science TV broadcast:

> Perhaps I can best describe my experience of doing mathematics in terms of a journey through a dark unexplored mansion. You enter the first room of the mansion and it's completely dark. You stumble around bumping into the furniture, but gradually you learn where each piece of furniture is. Finally after six months or so, you find the light switch, you turn it on, and suddenly it's all illuminated. You can see exactly where you were. Then you move into the next room and spend another six months in the dark. So each of these

breakthroughs, while sometimes they're momentary, sometimes over a period of a day or two, they are the culmination of —and couldn't exist without— the many months of stumbling around in the dark that precede them. (see e.g. Byers 2007, p. 1)

Now I quote more extensively from David Ruelle and Alain Connes. To make their reports more easily digestible to academic scholarship, I provide interlined comments in italics to their texts. Ruelle's and Connes' testimonies are our *textes d'explication*. Ruelle is a member of the *Institut des Hautes Études Scientifiques*, a colleague of Alexandre Grothendieck, Jean Dieudonné, and René Thom. He has made major contributions to dynamical systems and chaos theory.

> Mathematical research requires mental agility and the patience to pace around an infinite and dreary logical labyrinth *to examine and contemplate a connected series of mathematical concepts and images* until you find something that has not been understood before you: a new point of view, a new proof, a new theorem. (Ruelle 2007, p. 79, italics added)
>
> Doing mathematics is thus working on the construction of some mathematical object and resembles other creative enterprises. (op.cit., p. 88)
>
> The set of tools available to a mathematician may be compared to the system of highways available to a traveler; both provide the means to go efficiently from A to B. But [...] the highway system reflects the geography of a country, which we also know by other methods, so that building another road will not significantly change our knowledge of the geography. The panoply of technical tools of mathematics reflects the inside structure of mathematics and is basically all we know about this inside structure, so that building a new theory may change the way we understand the structural relations of different parts of mathematics. (op.cit., p. 94)
>
> *[In studying mathematical concepts we make use of appropriate tools, devices, and methods, which themselves are also mathematical concepts, and are studied as such. Improving or revising these tools changes the possibilities and relations of our network of mathematical concepts.]*
>
> Putting together a sequence of mathematical ideas is like taking a walk in infinite dimensions, getting from one idea to the

next. And the fact that the ideas have to fit together means that each stage in your walk presents you with a new variety of possibilities, among which you have to choose. You are in a labyrinth, an infinite-dimensional labyrinth. The ideas are human, and they belong to a human mathematical culture, but they are also very much constrained by the logical structure of the subject. The infinite labyrinth of mathematics has thus the dual character of human construction and logical necessity. And this endows the labyrinth with a strange beauty ... only through long study do we come to taste fully the subtle and powerful aesthetic appeal of mathematical theories. (op.cit., p. 96)

[The "labyrinth" is simply a suggestive name for the complex multiply-connected network of concepts and methods available for consideration and contemplation by the mathematician.]

Constructing a mathematical theory is the essence of mathematical work. [...] Constructing a mathematical theory is thus guessing a web of ideas, and then progressively strengthening and modifying the web until it is logically unassailable. Before that point you don't have a theory. In fact, it is usually not assured at the beginning that you will be able to complete your construction as originally planned (otherwise, the theory would be uninteresting). Clearly, during your construction work, you should concentrate your effort on the more uncertain links in your arguments. This is where your theory is most likely to fail, and you save time by knowing this early on. The easy and safe steps are left for later and are often handled in the final write-up by a dismissive sentence: "it is obvious that ...", "it is well-known that ...". (op.cit., p. 114)

Doing mathematics is often an individual and solitary enterprise. But mathematics as a whole is a collective achievement. (op.cit., p. 108)

[The mathematician thinks about a problem or searches for a new idea. This means examining the concepts he possesses, using the tools he has available from established mathematics. These concepts and tools are shared with the rest of the mathematical community. There may be some that he himself has invented and which are presently known only to himself.]

A mathematician lives in an intellectual landscape of definitions, methods, and results, and has greater or lesser knowledge of this landscape. With this knowledge, new mathematics is produced, and this invention changes more or less significantly the existing landscape of mathematics. (op.cit., p. 108)

[The term "landscape" is metaphorical, and can be replaced by "mathematical world".]

A mathematician who has finally understood a question may say that it was after all very simple. But this is usually an erroneous feeling. In fact, when our mathematician starts writing things down, their complexity unfolds and may end up looking formidable. A simple mathematical argument, like a simple English sentence, often makes sense only against a huge contextual background. (op.cit., p. 87)

And finally, quotes from an article by Alain Connes in the monumental compendium created by Timothy Gowers, *The Princeton Companion to Mathematics*. Connes is outstanding even among Fields Medal winners. His non-commutative geometry is a vast continuation of John von Neumann's work on operator theory.

The scientific life of mathematicians can be pictured as an exploration of the geography of the "mathematical reality" which they unveil gradually in their own private mental home. (Connes 2008, p. 1006)

[The trained and educated mathematician possesses a huge collection of mathematical concepts of all sorts from established mathematics (definitions, theorems, diagrams, graphs, problems both solved and unsolved) and modes of reasoning about them (standard arguments and calculations, heuristics including analogy and induction). Purposeful thinking about a specific problem, or openly seeking something interesting and worth pursuing, develops a familiarity with the connections and relationships of these concepts, which is analogous to the geography of one's neighborhood in some town or city.]

Once a mathematician truly gets to know, in an original and "personal" manner, some small part of the mathematical world, however esoteric it may look at first, the journey can properly start. It is of course vital not to break the "fil d'Ariadne" ("Ariadne's thread"); that way one can constantly keep a fresh eye on whatever one encounters along the way, but one can also go back to the source if one ever begins to feel lost. (loc.cit.)

[In this mathematical exploration, the mathematician may be led into "strange territory" (a domain of mathematics with which he is not yet thoroughly familiar). Connes advises him in that case, to look back at the starting point where he was thoroughly "at home" (i.e., completely familiar and in mastery of the concepts and tools).]

It is also vital to keep moving. Otherwise, one risks confining oneself to a relatively small area of extreme technical specialization, thereby limiting one's perception of the mathematical world and of its huge, even bewildering, diversity. (loc.cit.)

[This is advice which some mathematicians follow and some do not.]

The fundamental point in this respect is that even though many mathematicians have spent their lives exploring different parts of that world, with different perspectives, they all agree on its contours and interconnections. (loc.cit.)

[The internal images, representations or models possessed by different mathematicians, while private and idiosyncratic, are "congruent", that is, they give the same answers to test questions about "interconnections". They match each other in the ways necessary to permit mathematicians to talk to each other about these concepts with understanding and agreement.]

Whatever the origin of one's journey, one day, if one walks far enough, one is bound to stumble on a well known town: for instance, elliptic functions, modular forms, or zeta functions. (loc.cit.)

[This is a testimony to Connes' own experience, or a statement of faith in mathematical thinking in general. If one hasn't stumbled on a well known town yet, then one tells oneself that one simply hasn't walked far enough yet.]

It would be easy to make a translation between Connes' account (traveling through a mathematical landscape) and that of Wiles (exploring a darkened room). Connes' image of a landscape is similar to

Alexandre Grothendieck's description, in *Récoltes et Semailles*, of his feelings on leaving functional analysis for algebraic geometry: "I still remember this strong impression (completely subjective of course), as if I was leaving dry and gloomy steppes and finding myself suddenly in a sort of 'promised land' of luxuriant richness, which spread out to infinity wherever one might wish to put out one's hand to gather from it or delve about in it" (quoted in Connes 2008, p. 1006).

What does all this add up to, Grothendieck, Thurston, Newton, Wiles, Connes, Schutzenberger, Ruelle, Harris and Heintz?

With different metaphors and each in their own way, they tell us that mathematical work feels like direct contact, bumping into actual objects, which the mathematician manipulates, "fools around with", plays with, turns over, tries to connect to other mental objects, until finally he/she "sees" what is going on.

To Andrew Wiles, it is like stumbling around an unfamiliar room and banging against the furniture, until at last a light goes on and he sees clearly where he is. (Then he is ready to enter the next dark room.) David Ruelle and Alain Connes talk of mathematics as a landscape, or as a labyrinth through which one travels and searches. Such testimonies have not usually been taken seriously by philosophers of mathematics. But they are not lunatic ravings, nor are they senseless inexplicable poetry. They are important evidence, which needs to be interpreted.

These authors know what they are talking about. Their accounts are not fanciful fiction. They are autobiographical reports of actual experience. Anyone interested in how mathematics works would be mistaken to dismiss or ignore them.

Of course there is no "furniture" or "labyrinth" in any literal sense. What there is, is actual mathematical thinking! Wiles and Ruelle and Connes are encountering mental mathematical entities. These mental mathematical entities are experienced as actual objects, more or less clearly or obscurely perceived, that have their own properties, which the mathematician may struggle for a long time to ascertain.

This is psychology, or "psychologism", long mocked and outlawed in the philosophy of mathematics. And it is "introspection", long ago outlawed as unscientific in the world of "scientific psychology". Nevertheless, there can be no substitute for such self-reports of actual practitioners. Yes, do also analyze their published research articles, to see what is present there. But what is not present there, that is the very essence of the matter.

There are two mistaken ways to respond to this internal reality. A naïve materialist minimizes or denies it. ("Thoughts and concepts are ephemeral, immaterial. Rocks or stars or neurons, what is detectable by scientific instruments, *that* is really real.") On the other hand, a mathematical Platonist mythologizes it, spiritualizes it, makes it superhuman and transcendental (It is "out there".)

Wait! Someone raises an objection.

"The mathematical thinking described by Ruelle and Connes and Wiles—isn't that just heuristics, discovery? Isn't that a different matter from demonstrative reasoning, from *proof*, where formal logic is relevant?"

Answer: "Demonstrative reasoning" is the reasoning by which one *follows* step by step "along the ridge" leading to the "peak", the mathematical result. You follow the proof by means of the properties of mathematical entities, properties which are either directly observed from your mental model, or else referred to in the literature. It is not done syntactically, it is not done by transforming formal sentences.

First we have to find "the ridge leading to the cliff", the steps by which to lead our reader to see our result-the proof! The *problem* that Wiles was solving was to *find a proof*. He was stumbling around in the dark, looking for *a proof* of Fermat's conjecture. Certainly, a big part of mathematicians' work is forming conjectures, looking around for patterns. But the biggest effort in our research is in *finding proofs*. After I have a conjecture, which I found heuristically, finding a non-routine proof again requires heuristics. After *I* see the proof, I have to organize it, present it coherently, in a way that my reader or listener can follow. Some mathematicians have found this very last step to be the greatest obstacle.

4 The Materialist versus the Platonist: Changeux and Connes

The materialist met the Platonist in *Conversations on Mind, Matter, and Mathematics* between the neuroscientist Jean-Pierre Changeux and the mathematician Alain Connes. (Changeux and Connes 1998; a critique of the *Conversations* can be found in Petitot 2005.) The materialist viewpoint is clearly expressed by Changeux (see also Behrens 2012). He says: "Mathematical objects exist materially in your brain. You examine them inwardly by a conscious process in the physiological sense of the term. Because these objects have a material reality, it's possible to study their properties. In the head of a mathematician, mathematical objects are material objects —mental objects if you like— with properties that are analyzable by a reflexive process. It ought in principle to be possible to observe them from the outside looking in, using various methods of brain imaging" (Changeux and Connes 1998, p. 12).

Observe mental objects by brain imaging? Can anyone, even in principle, by "various methods of brain imaging" observe, for example, the commutative law of multiplication, inside my brain? The commutative law is a permission for me to make a certain operation or transformation in an arithmetical calculation. By electro-chemical methods of observation, we observe electro-chemical phenomena. An electro-chemical phenomenon is not a law of arithmetic. The commutative law of multiplication simply is not of the same "category" as any physically observable electro-chemical process.

Yet our mathematical thinking, like any kind of thinking, conscious or unconscious, IS based in or rooted in our bodies (mainly our nervous systems).[2] So our mathematical concepts in some manner or representation are inseparable from brain components or activities—electro-chemical entities. To the mathematician's *mental struggle* to solve a problem, taking place "in his head", there must be a corresponding *brain process*, an electro-chemical process, on the bio-physical level. Nieder (2011) is an impressive example of work to reveal this connection. But impressive as is research such as Nieder's, it is not directly relevant to the mathematician or the philosopher. When we are doing mathematics—creating, elaborating, and connecting mathematical objects, we can do so only by dealing with them as having content, meaning something. Thinking of mathematical concepts as merely properties of neurons would make it impossible to do mathematics.

Alain Connes, Changeux' opponent, is a forthright Platonist. "There exists, independently of the human mind, a raw and immutable mathematical reality, and on the other hand, we have access to it only by means of our brain at the price of a rare mixture of concentration and desire" (Changeux and Connes 1998, p. 26). [...] "My position cannot change ... it's humility finally that forces me to admit that the mathematical world exists independently of the manner in which we apprehend it" (op.cit, p. 38). Connes certainly knows his own mental mathematical experience. It would be absurd to doubt his reports of his inner mental process. Many another mathematician has said much the same thing: "Our knowledge of mathematical reality is certain, it is unquestionable!"

But we have to distinguish two aspects of such claims. There are reports of actual experience when doing mathematical work, and there

---

[2]Need a proof? Mathematical work often involves your eyes, to read something, or your fingers, to write something. Signals go from your eyes up through your optic nerves into your brain (into your mind). Other signals go down from your brain through other nerves into your fingers (according to your mental decision what to write). Your thoughts, feelings, and inner consciousness, your mind, somehow is "there", in your nervous system and your brain (although in ways we may never in detail understand).

are interpretations of actual experience. What about the claim that the truths of arithmetic are eternal and unchanging? Such a claim cannot, on the face of it, be based on direct experience. It is a statement of belief, not a report of direct experience. What of the claim that these truths are independent of the knowledge or the existence of human beings? Such a claim is also, on the face of it, not something that one could know from direct experience.

Connes' subjective experience of doing mathematics, which is experienced in common by many mathematicians, and described by several of us, is not a delusion, nor a superstition, nor a myth, any more than the inner, personal, subjective experience of people in general. He experiences mathematical concepts directly in working with them as a researcher. He cannot be swayed from his conviction that they are real, and independent of his own awareness of them. That opinion of his is perfectly correct. These concepts are embedded in the vast intertwined structure of living mathematics, which is a social structure shared by the mathematical community as a whole. It exists independently of the awareness of any individual mathematician.

Bur for Connes, it is necessary to describe this objectivity in the traditional language of Platonism. "Mathematics is absolute, universal, and therefore independent of any cultural influence" (op.cit., p. 50). It is "out there", he says, neither in space nor in time, eternal, unchanging, immaterial, extra-human. This claim is an interpretation of his experience, not an experience or an observation. It is neither verifiable nor refutable. Anyone is free to believe it or not. And it is incompatible with ordinary scientific discourse, which long ago rejected the dualism of separate incomparable "Substances" called "Spirit" and "Matter".

When he insists that there are facts of the matter, which are what they are, regardless of our desires, Connes is correct, and entirely justified. These facts are properties or aspects of our shared mental models. Connes' attraction to Platonism, which is shared by many other mathematicians, expresses our conviction that we are dealing with something real, something that goes beyond our own individual awareness. But there is no necessity to place it beyond time, space, and human consciousness.

Connes and other mathematicians observe our inner lives as we work mathematically, and then reflect and recall about such work. In order to talk about it, we resort to metaphors—a "darkened room", a "labyrinth", a "landscape". Of course there is no darkened room inside Andrew Wiles' head, no labyrinth inside David Ruelle's! However, our inner world of mathematical knowledge and thought is an objective reality. We explore it and try to understand it. We navigate it, not arbitrarily, or as we might hope or prefer, but according to its own possibilities. When we invent or create new mathematical entities, they too become parts of this inner world of objective reality, with their own properties, some that are clear and apparent, and others that are hidden and inscrutable.

So who is right, Changeux or Connes? Are mathematical concepts the thoughts of Alain Connes, or are they electro-chemical traces in Connes' brain? Both Connes and Changeux are right. Changeux is right about the neurons and the brain, and Connes is right about what he actually does when he does mathematics. They are not contradictory! They are complementary. Changeux and Connes are talking past each other. The facts that Connes is emphasizing are not inconsistent with the facts emphasized by Changeux. They report two different ways of observing the same thing, from two radically different points of view.

The disagreement is in Changeux' implied "nothing but". Mathematical concepts are properties of the brain, but they are not nothing but properties of the brain. Connes parries Changeux' reductionism, saying "It might also be an illusion—like believing that if we only knew more about the chemistry of ink and paper we would have a better understanding of the works of Shakespeare" (op.cit., p. 14). The seeming contradiction is created by the reductionist impulse—the impulse to say "either-or", rather than "both-and".

To see how two such different pictures can fail to be contradictory, it is enlightening to compare the Changeux-Connes conversation to an ordinary conversation between a physician and a patient. The patient reports "light-headedness", "fatigue", "confusion". The physician considers: "Auto-immune reaction?", "Infection?", "Head injury?" ... No one would ask who is right, the physician or the patient? Everyone understands that both aspects are real, and complementary! Both the patient's inner subjective report, and the physician's external, physical explanation.

How does a physician know what a patient means by "light-headed" or "confused" or "fatigued"? He/she knows, because he/she also can experience such sensations, because he/she also possesses a human body and brain. The patient is in the physician's "lifeworld", to steal a phrase from Husserl or Habermas. A common biological and cultural lifeworld. And how do mathematicians understand each other, about their inner thoughts, guesses and insights? Because they too share a "lifeworld". They are all human beings, first of all. And on that basis, their common training and education brings them to their common understanding.

Here is another metaphor. Think of a huge cathedral, like Notre Dame de Paris. One observer is locked outside of it. He sees turrets and gargoyles, but he has no access to the interior. A second observer is locked inside. She sees glorious statuary and gold ornaments, but she has no access to the exterior. They communicate "on line". But they cannot understand each other. Their two descriptions seem to be utterly contradictory. But both are correct! What's lacking is the insight, that their two descriptions are complementary.

Consider two complementary views of a spherical surface like an eggshell. From the inside, it is concave. From the outside, it is convex. Which is correct? Both, of course! Mathematical entities (objects, concepts, theorems, algorithms, problems, conjectures, analogies, and so on) are things that we think about. So they are available for us to think about. They are mental entities. And also, at the same time, in some form or realization, they are physically present in our brains. So they have definite properties of two kinds—both mental, as shared mathematical activities, and physical, in our brains.

The puzzle about mathematical proof is dissolved if we realize that mathematical proof is about a kind of mental object accessible to mental inspection. By thinking about numbers (or functions or operators), we can sometimes come to understand, to see, that a certain statement about numbers is correct, is "actually the case". Then we tell others how they too can come to see that fact about numbers.

We aren't able to show this by the methods of neuroscience. But it's not surprising that mental models —which are also brain activities— can be directly observed mentally—which is to say, by the brain itself. That "observing", "seeing", is what mathematicians mean by a "proof". That's what Hardy, Ruelle, Connes and Wiles are trying to tell us.

I use the expression "mental model" for the internal entity in the mind of anyone, including a mathematician, any entity, object, or process that one may think about, concentrate on, study by inner thought. A mathematical concept is a collection of mental models that are "mutually congruent", fit together. The concept "triangle", for example, is a shared, public, inter-subjective entity. Each of us who "understands" the word "triangle" has his/her own internal entity, available for contemplation or mental manipulation. That inner, private mental entity corresponding to the shared concept is what I mean by our "mental model". Under the pressure of a strong desire or need to solve a specific problem, we assemble a mental model which the mind-brain can manipulate or analyze.[3]

The mathematician's inner life, even though reported by metaphor,

---

[3] The term "model" seems to be the most appropriate, even though it risks misunderstanding. The notion of "mathematical modeling" in applied mathematics is not what I mean. The psychologist P. N. Johnson-Laird has used the term in a specialized way, not what I mean (Johnson-Laird 1983).

is a fact, not a fiction. The inner life of the mathematician is just as real as yours, dear reader! The philosophy of mathematical practice cannot answer some of its leading questions without recognizing and dealing with this reality—the subjectivity or inner life of the mathematician.

While mathematical thinking, like other thinking, is in an important sense private, even perhaps incommunicable, yet in another sense it is public and open. Mathematicians share a culture, a system of knowledge, a tradition, a set of methodologies, and a special kind of subject matter that permits them a unique and very precious *consensus*, even *unanimity*. In other kinds of theories, in "the humanities", theorists may freely continue to disagree for centuries. But in mathematics, with rare and brief exceptions, once a new mathematical result is proved —once a "proof" has been accepted by the appropriate referees, editors, and experts— then it becomes part of *established mathematics*. Then every mathematician is free to use it as a tool or a building block in creating or discovering new mathematics.

This testimony is subjective, or introspective, but that does not mean it is private or impenetrable. On the contrary, the essential characteristic of mathematical thinking is that it is shared by the mathematical community. *Our training, our education, and our work experience* make our personal mental models congruent, fitting in with each other. The social life of mathematics makes these personal, internal mathematical objects "mutually congruent", or matching.

It is helpful to borrow some terminology from mathematics itself. "Equivalence class" is a fundamental elementary mathematical notion. It denotes a set of things that are equivalent to each other in some definite sense. The members of an equivalence class are individuals who share certain properties with other members of the class. Such an equivalence class itself becomes a well-defined entity, with its own definite properties. For instance, the triangles of all shapes and sizes, that mathematicians talk about, are each members of the equivalence class, or concept, "triangle". The numeral "3" is an equivalence class. All the various symbols for "3" that are actually written on paper or blackboards are "representatives" of this equivalence class. They are not identical, but they all serve to represent the "class", which *is* the numeral "3" (and which, on this page that you are now looking at, is represented by whatever representatives my computer program generates).

In our present context, dealing with mathematical cognition —thinking, proving, and problem-solving— we talk about the various individual mathematicians' *mental models* of some mathematical concept. All these individual mental models of the concept are equivalent—that is, mutually interpretable, communicable, with agreement and understanding. Together, they form a *collection* of mental models which are equivalent to each other in that sense, of being mutually interpretable, communicable, with agreement and understanding. We can think of them as a kind of equivalence class, like a mathematical equivalence class.

In effect, this equivalence class itself is the "mathematical concept". My interior mathematical objects are personal and subjective, but the "equivalence class" of mutually congruent models is inter-subjective, it is cultural. *My personal representatives of the concepts are internally available for manipulation and experimentation. In that sense, my knowledge of mathematical entities comes from direct perception.*

The "proof" is a procedure, an argument, a series of claims, that every qualified expert understands and accepts. It is the possibility and necessity of proof that defines mathematics. (To the extent that physics or linguistics or genetics have proofs, we speak of *mathematical* physics, linguistics or genetics.) But who are these "qualified experts"? Anyone who understands the concepts involved in the proof! That is to say, anyone who has acquired or constructed the necessary mental models. In some cases (like Wiles' proof of Fermat's Last Theorem) only a small group of leading specialists are thus qualified. In other cases, which use only high-school level mathematics, you, the reader, are qualified. (This paper concludes with my example of a mathematicians' proof accessible to most people.)

The mental-social reality of mathematics is different from other socially shared beliefs, in crucial ways. It has persisted for centuries, and grown to a vast extent. Established mathematics is universally accepted and adopted, in every part of the world. This impressive fact calls for an explanation. It is plausible that mathematical thinking is a general aspect or property of the human brain—not an innate *ability* everyone is born with, but a *capacity for development* that every sane adult possesses to one degree or another, some only slightly, others very highly.

Bodily and visual sensing of distance and direction, from which *geometry* arises, are shared by us with every mammal that is ever either a predator or a prey. And the discovery of "subitizing" in weeks-old infants seems to establish an inborn basis for *counting*. Not such a novel idea, really. *Old Kant was right* when he said that mathematical ability is based on something innate in our mind-brain. His mistake, to fetishize Euclidean geometry, is only a detail. Even Plato was right, when rightly interpreted, in claiming that geometric reasoning is remembered from before we were born. He put it up in Heaven. We put it rather in evolution, in genetics.[4]

There remains a gigantic scientific challenge: to refute or to verify these claims, by detailed, specific knowledge of the brain and nervous system. Some beginnings of such knowledge have accumulated.[5] We might wish for specific, refined knowledge, such as: "Exactly what electro-chemical circuits in the brain of Isaac Newton or Terry Tao made their discoveries possible"? Such knowledge may forever remain out of reach.

The mathematician has direct access to her own representatives of the concepts that she shares with the mathematical community, which are part of *established mathematics*. I call these representatives her "mental models", to bring out the compatibility of this statement with contemporary neuroscience. The neural description of thought and the psychological or introspective description of thought are complementary descriptions of the same thing. By reminding you that these mathematical concepts have a material basis, I hope to convince you that they are real, objectively existing entities, even though they are accessible to us only in our thoughts.

5 What, then, is a mathematicians' proof?

Mathematicians' proofs compel agreement and acceptance by leading the reader along a path in his/her mental mathematical universe, to where he/she "sees" the claimed result—or more conventionally stated, "sees" that it is "true" or "correct".

And when must such a proposed proof be rejected? Since the proof-claims that are accepted and the ones that are rejected both are deficient as logic, what makes a proof-claim unacceptable? Simply, that it fails. Meaning simply that the reader is not led to "see" the proposed result.

But does "proving" a mathematical claim, in the sense of mathematicians' proof, guarantee that it is "true"? There is no need to step into the notorious quagmire of trying to define "mathematical truth". The history of mathematics shows that mathematics undergoes a continuing process of self-correction and clarification. Is Euclid's parallel postulate "true"? In one sense yes, in another sense no. Euclidean geometry is a well-established theory, a collection of well-established theorems, that mathematicians are able to use, to create and establish new mathematics.

"*Anecdotal description is all very well,*" someone might object, "*but*

---

[4] Yehuda Rav has written incisively about the evolutionary basis of mathematical thinking. See Rav (2006).

[5] See Dehaene and Brannon (2011).

what about Right and Wrong? Don't we need logic, to tell us what we **ought** to do?"

The mathematician using established mathematics is comparable to a normal human being using vision and hearing. While our eyes and ears may deceive us, we must trust our eyes and ears (using due precautions, of course). Refusing to use them would be insane. By using our eyes and ears, we stay alive; if I refused to use them, I would die.

If I want to do mathematics, I have to acquire and use the concepts in established mathematics, the concepts shared by the mathematical community. That is what doing mathematics means. Acquiring those concepts means acquiring my own mental representatives of them, my own mental models.

To the question, *"What qualifies the usual informal mathematical arguments as 'proof'?"* I can now answer, in two parts. First of all, the usual informal mathematical arguments are accepted as "mathematicians' proof" if they convince readers who are qualified—who possess the appropriate mental models. This "convincing" is based on observing —directly verifying— that their own mental models have the properties claimed in the "proof". This is the way new results are added to established mathematics.

Secondly, the question then arises. What "qualifies" established mathematics itself? Some have tried to answer this, by providing mathematics with a *foundation*. But such work, interesting and fruitful though it may be, leaves us asking, what "qualifies" the proposed "foundation"? With or without "foundations", the world's businesses, budgets, machinery and technology operate on the basis of established mathematics. As many have already said, *this* in the end is what "qualifies" it. Mathematics is part of life, which does not have to be qualified.

When we do "applied mathematics", we relate mathematical entities to physical ones. But even then, it would usually be wrong and misleading to think of the mathematical entity as being *in the first place* a representation of a physical one. This fact is obvious for 4- (or 5- or 6-) dimensional geometry, since ordinary physical space is limited to 3 or fewer dimensions. The n-dimensional hypersphere is a mathematical entity in the minds of individual mathematicians. All mathematicians acquainted with the n-sphere will give the same answers to test questions about it. True, the n-sphere can be defined axiomatically, but that possibility is simply one part of the information contained in the mathematical model. Mathematicians' command of the n-sphere is based largely on intimate acquaintance with the special cases $n=1$ and 2, together with a refined sense of how parts of that knowledge can or cannot be carried over to higher cases. In physics, there are "phase spaces", which we can think of in terms of four or more-dimensional geometry. But the study of four or higher dimensional space does not depend on such interpretations.

To complete a mathematicians' proof, one usually must also include calculations. Any special field of mathematics has its associated calculations, which are often done by machine, using programs such as *Matlab* or *Mathematica*. A mathematical discovery or proof often involves *an insight that a certain fact can be discovered or verified by a certain calculation*. Then one does the calculation, to see if the insight is borne out or rejected. But to interpret the calculations in mathematicians' proof as meaning that such proofs are essentially symbol-pushing would be an elementary and fatal mistake. Indeed, one of the gravest signs that an aspiring mathematics student is out of his depth, and may be in the wrong classroom, is being caught thinking or talking *formally*: concentrating on the symbols, rather than *interpreting* the symbolism as *representing* concepts. Mathematical thinking is conceptual thinking. The symbols can be transformed or replaced while the meaning remains the same.

6 Relation between formal proof and mathematicians' proof

The heritage of Leibniz, Frege and Russell persists, in the notion that mathematical proof is, can be, or should be reduced or reducible to syntactics (first-order logic). Impressive formalizations of significant parts of mathematics have actually been achieved, showing that the formalization of significant parts of mathematics is actually possible.[6] To the extent that such formalizations are mathematical projects, they must be carried out by the "usual, informal" kind of mathematical reasoning.

Logic and mathematics are two distinct strands of rational thinking. Formal logic is sometimes thought of as a *model* of how mathematicians think, and on other occasions as a guide to how they *ought* to think. Simultaneously, mathematicians recognize mathematical logic as "another branch of mathematics, like geometry or number theory", with practitioners thinking in the "usual, informal" mode of mathematicians. The two distinct strands are doubly intertwined. Indeed, their mutual fertilization is possible precisely because they are distinct.

When it is mentioned that mathematicians' "proof" is not formal proof, the discrepancy is sometimes explained by saying that mathematicians' proof is *an abbreviation* of a formal proof. "For convenience and readability, mathematicians find it advisable to leave out logically necessary steps in their proofs". So a mathematicians' proof is then sup-

---

[6]See Hales (2008) and Gonthier (2008).

posed to be *a token or a promise* of a full logical Proof. Somebody could fill in the gaps—turn the mathematicians' proof into a formal proof.

The mathematician, in most cases, has neither the knowledge, the ability, the interest, nor the willingness, to do anything of the kind. And our logically defective proofs actually do what they are meant to do: convince our fellow mathematicians!

If and when such a formalization is actually carried out, something new and interesting may be learned. Perhaps the formal proof demands a lemma that was not stated in the original proof. A new specialty in logic, "proof mining", seeks such improvements. Nevertheless, the possibility of such a formalized syntactic version is not what makes mathematicians' proofs convincing to mathematicians.

"Well then," the logician may rightfully ask, "let the mathematician tell me what he or she means by proof!" A reasonable request. Since it isn't Proof as Proof is known in logic, how does it serve the purpose of proof—namely, to compel agreement?

"The working mathematician can be likened to an explorer who sets out to discover the world. One discovers basic facts from experience. We run up against a reality that is every bit as incontestable as physical reality" (Connes in Changeux and Connes 1998, p. 12).

What is this "reality"? It is our internal concepts, our mental models, which are *real objects, with real properties*, and which are congruent to each other, which fit together and match. That's the sense in which it is "the same thing" that we are looking at. The term "mental model" is meant to suggest substantial existence or autonomy.[7] The common term "representation" is misleading, because it suggests that some other more substantial or even physical entity is being "represented".

In a different sense of the word "model", in the usual sense of applied mathematics, the formal proof is an important "model" of mathematical proof. This sense of the word "model" is different from that in the rest of the present article. Here, as usual in applied mathematics, "model" means a mathematical structure that has an interpretation in terms of a certain "real-world" phenomenon. In this instance, using formal proof as a "model" for mathematicians' proof, the formal proof plays the role of the mathematical structure! The "real-world" phenomenon is mathematics itself.

The zero'th principle of mathematical modeling is, DO NOT confuse or identify the model with the process being modeled. Fluid dynamics is not water. We study water waves by means of fluid dynamics, but that

---

[7]The notion of the mathematician's "mental model" was amplified in Hersh (2011).

is no substitute for diving and swimming!

One may use formal proof as a "model" of mathematicians' proof, but mathematicians' proof is not formal proof. Our starting point is established mathematics, not some postulated axioms, and our reasoning is "semantic", based on the properties of mathematical entities, rather than "syntactic", based on properties of formal sentences. Our testimony may be called "anecdotal", but it is not fictitious, it is lived experience. It is real evidence.

The formal logic model of mathematics is a great success. The model of mathematics as a formal system has made possible many amazing insights—going far beyond its monumental beginnings (Gödel's incompleteness theorem and Turing's unsolvability theorem). But one may be interested in mathematics itself, as well as in its models, even if such an interest does not produce deep theorems or powerful technologies.

Many mathematicians testify that mathematical knowledge is strictly analogous to the empirical scientist's knowledge of his/her objects of study. Connes and Schutzenberger plainly say that they see no difference between themselves, studying mathematical concepts, and entomologists, studying insects. (Connes, Lichnerowicz, and Schützenberger 2001)

Cognitive neuroscience has a different way of describing and studying the objects of thought, locating them in the nervous system and the brain. Our subjective or introspective testimony is not competing with or contradicting the testimony of neuroscience. It is complementary to it. These two different kinds of descriptions are complementary ways of reporting on our "internal mathematical objects", or "mental mathematical models".

7 Aristotle, Kant, and Locke

Mathematician's proof is based on a kind of direct *seeing*—internal "seeing", of course, rather than external. This is not a novel or unfamiliar opinion. Wedberg (1955, pp. 88–89) cites both Aristotle and Kant:

> Aristotle emphatically asserts that the geometrical figures have only a potential existence before they are brought to actuality through the geometer's thinking ... The geometer's thinking is an actuality ... it is by making constructions that people come to know them. (Metaphysics 1051 a 21-33)
>
> As soon as the appropriate construction has come into existence, the construction itself is directly seen and the proposition to be proved is immediately understood to be true.

[Not usually that the proposition to be proved is immediately understood to be true, but rather, that each step in the argument is immediately seen to be true, from the preceding step. But still, it is SEEN to be true, not syntactically deduced to be true.]

Aristotle seemingly anticipates some of Kant's most characteristic views, viz. (a) that the geometer carries out constructions in an intuitively given space, and (b) that in establishing geometrical theorems the geometer makes essential use both of logical deduction from axioms and direct inspection of his construction in the intuitively given space.

John Locke is another respected philosopher who had similar views (see the quotations in Hersh 1997). These references to Aristotle, Kant and Locke show that what I am saying is not novel or unprecedented among philosophers. The evidence and argument in this article is making two points. First, mathematical concepts are neither fictional nor transcendental, they are real mental entities, with definite properties of which we can have reliable knowledge. Second, mathematicians' proof ("informal proof", as some would have it) *works*, it is *compelling*, because it uses direct observations of mental entities *accessible to the mathematician*.

Is this mere Platonism? With Platonism, it shares the assertion that mathematicians directly perceive properties of mathematical objects, but WITHOUT giving those objects any superhuman, transcendental or eternal existence "out there". They are "down here", in the shared, or collective, or public consciousness of thinkers, of us humans.

Recognizing and accepting that mathematician's concepts are real entities (as thoughts in the minds of individual mathematicians, and as the shared equivalence class of such models), will help clarify an old puzzle about standard and nonstandard models. The puzzle arises when a formal axiomatic system has several mathematically distinct interpretations. The formal system can't tell which interpretation is the "standard" or "intended" one. But if we step away from the formal axioms, we can say that the standard or intended interpretation is just the one that is actually present in the thinking of mathematicians, whether or not one is able to characterize it axiomatically.

In arithmetic, a nonstandard interpretation of the Dedekind-Peano axioms was discovered by the Norwegian logician Thoralf Skolem. In set theory, Paul Cohen created non-standard interpretations of the Zermelo-Fraenkel axioms. Cohen viewed accepting or rejecting the Continuum Hypothesis (CH) as a matter of choice or preference. For the present time, the standard interpretation is to leave CH undecided. On the other hand, Hugh Woodin is working to decide the correct choice, based on its set theoretic consequences.

8 Heron's area theorem

I will take as an example a simple modern derivation of a classical formula for the area of a triangle, as a function of the lengths of the three sides. It is called Heron's formula, but may actually be due to Archimedes. Although it is simple and useful, it's not in Euclid, and it isn't taught in standard high-school geometry. The traditional proof relies on similar-triangles constructions that are long and tricky. Dunham (1990) gives a very nice presentation. He calls the formula "surprising" and "un-intuitive", because it involves a square root operation, and an unfamiliar variable s, the semi-perimeter (half the sum of the lengths of the sides). I will present an easy derivation of Heron's formula, using simple high-school algebra, in order to illustrate the points I have been making about mathematical thinking and mathematicians' proof.[8]

As presented in standard references, Heron's formula gives the area of a triangle with sides of length a, b and c as the square root of

$$s(s-a)(s-b)(s-c)$$

s is the "semi-perimeter", $(a+b+c)/2$. In the proof attributed to Heron, s is the radius of a certain circle essential for the proof.

When I read the proof in Dunham's book, it seemed to me unreasonable to work so hard and be so tricky, in order to prove something so simple.

We learn in 10th grade that triangles with equal corresponding sides are congruent to each other, so of course they have the same area. Therefore, the area is a function of the three side-lengths a, b, and c. Why should the formula involve the irrelevant quantity, $s = (a+b+c)/2$?

This objection to the standard Heron formula is esthetic. Yet it will be understood and accepted by any mathematician.

So we replace s by $(a+b+c)/2$ and simplify. There now appears under the square root sign a new product of four factors:

$$[(a+b+c)/2][(-a+b+c)/2][(a-b+c)/2][(a+b-c)/2]$$

This expression is more natural and appealing. The three variables a, b, and c are all treated alike. That's as it should be, since the area

---

[8] This presentation is similar to the one in Hersh (2007). The same idea had previously been published by R. C. Alperin as a classroom note (Alperin 1987).

doesn't depend on how we label the three side lengths. (This remark, introducing symmetry reasoning, while completely elementary, goes beyond Euclid and introduces a modern viewpoint.)

In this formula, the three factors with minus signs are not what we would expect. They're like bumping against a piece of furniture. Where do they come from?

Let's visualize all sorts of triangles, acute and oblique, swimming around, changing size and shape as they swim around. Some of them degenerate, when a vertex approaches and collapses onto the opposite side.

Wait! Look at that! The area will be zero! When sides b and c collapse onto side a, and $a = b + c$, then the triangle degenerates to a line segment, and the area is zero!

Insight! A link between algebra and geometry! The area must be zero, if any side length equals the sum of the other two! Now pull in a little algebra from 11th grade high school—the Factor Theorem. If a polynomial in a variable a equals 0 when $a = b+c$, then the first degree expression $(-a + b + c)$ must be a factor of the polynomial. And by symmetry, the same must also be true for $(a - b + c)$ and $(a + b - c)$.

That's it! Almost everything is explained. If the area is a polynomial in the side length a (for example) then the expressions $(-a + b + c)$, $(a - b + c)$, and $(a + b - c)$ must be factors of the area formula, because when a equals $(b + c)$ or $(b - c)$ or $(-b + c)$, the area will be zero.

This little insight is like Wiles' light going on, making the furniture visible.

But wait a minute—that cannot be! We're back in the dark, with the furniture banging against our legs. The area of a triangle scales quadratically with length. (If you double all the sides of a triangle, for example, the area is multiplied by four.) The expression we were just imagining is a product of three first-degree factors, it has third degree, it would get multiplied by 8, not by 4. So it is wrong!

We have actually *proved* a little proposition, not deserving to be called a theorem: "To any polynomial in three variables a, b, c satisfying the triangle inequality (the sum of any two side lengths is greater than the third side length) there is a triangle with side lengths a, b, c whose area does NOT equal the value of that polynomial".

But how do we finish the derivation of Heron's formula?

It has to contain three distinct first-degree factors, but area cannot do so, because it is homogeneous quadratic. Then why not try the next best thing, the SQUARE of the area? That will be a homogeneous fourth-degree or "quartic" expression. We can get such a thing by just multiplying the three linear expressions we already have by one more first-degree expression.

That's just a guess, but it's the simplest one.

The first-degree expression has to be symmetric in a, b and c, so it can only be $(a+b+c)$, or maybe $(2a+2b+2c)$, or $(3a+3b+3c)$.

It is now clear that while the formula is NOT a polynomial, it COULD be the square root of a quartic, of the form

$$k\sqrt{[(a+b+c)(-a+b+c)(a-b+c)(a+b-c)]}$$

where k is some numerical constant. To determine the constant k, choose $a = b = 1$, and $c = \sqrt{2}$. This is a right triangle with area $1/2$. Our formula becomes

$$1/2 = k\sqrt{[(2+\sqrt{2})(\sqrt{2})(\sqrt{2})(2-\sqrt{2})]}$$

After simplifying we get

$$1/2 = k\sqrt{4},$$

so $k = 1/4$.

This is Heron's formula. We have *proved* that IF the area of an arbitrary triangle is the square root of a quartic function of the side lengths, then it is given by Heron's formula.

We have not yet proved Heron's formula completely, for we used an extra hypothesis. But once the formula is in our hands, a routine exercise finishes up, by Cartesian coordinates, or the law of cosines. The interesting part was *deriving* the formula. How did we accomplish that?

First we looked for the simplest formula, a polynomial in a, b, and c. The key insight was that the *geometric* condition for degeneracy —a vertex collapsing onto its opposite side— implies an *algebraic* condition: a certain first degree expression must appear as a factor in the area formula. By symmetry, the formula would then have to contain at least *three* first degree factors, and so would be of degree at least three. But we know that area is a *quadratic* scale function. Therefore we see that the formula CANNOT be a polynomial, and we *guess* the next simplest possibility—square root of a quartic. Using symmetry again, we see that only one quartic is possible, except for a constant factor, which we determine by choosing a convenient special triangle.

We used an insight connecting algebra to geometry, and then a geometry theorem, an algebra theorem, some plausible reasoning, and symmetry.

Most of this would be absent from anyone's list of axioms for elementary geometry. But all of it is available to an educated mathematician. Our derivation is convincing to anyone who understands the Factor Theorem, and the scaling of area, and reasoning by symmetry. This exercise is a little example of how mathematicians "prove". We do it by citing properties of the entities in question. Shared knowledge of those properties compels final agreement.

It would be easy to pad out this derivation formally, and then rewrite it syntactically. But to what purpose?

Instead, the mathematically natural next step is to generalize to three dimensions. Just as three non-collinear points in the plane define a triangle, four non-coplanar points in space define a tetrahedron, or triangular pyramid. What is the analog for the tetrahedron of Heron's formula for the area of a triangle?

A natural guess is a formula for the volume, as a function of the lengths of the six edges. It "should be" the square root of a sixth-degree symmetric polynomial in six variables. This natural conjecture FAILS! An account of how and why is given in Hersh (2007). The lesson is: a mathematical object is what it is, not what I might want it to be or imagine it to be! It may take a few bruises to the "knees and elbows" (speaking metaphorically) to recognize relevant properties or facts about this mental object.

9 Conclusions

This little excursion into plane geometry and high-school algebra is meant as an example, of the way mathematicians come to conclusions, and convince each other of their results.

The recognition that mental mathematical models or internal mathematical entities actually exist, are real things about which factual statements can be made, is the key to understanding mathematicians' proof. Mathematicians' proof depends on mathematicians possessing, recognizing and communicating congruent samples of certain equivalence classes of mental models. These equivalence classes of mental models are ordinarily referred to as "mathematical concepts". They correspond to brain processes which are congruent, or equivalent.

The claim that I am making is speculative. It asserts that certain things are going on in the brain or the nervous system, which we are not able to directly observe or verify. The claim is based on the testimony of mathematicians about what they experience, and on the plain fact: mathematicians *do* make discoveries about mathematical entities by just thinking about them, and these discoveries are ordinarily verified by other mathematicians by thinking about them (and also doing some calculations.)

Acknowledgements

Brendan Larvor, Vera John-Steiner, Edward Dunne, David Edwards, Colin McLarty, Martin Davis, Ulf Persson, Chandler Davis, Joe Auslander and Carlo Cellucci all provided essential sound advice and moral support.

Appendix

Following a reference in Timothy Gowers' blog, I came upon a wonderful entry in the online website called *Quora*.[9] On *Quora*, people post interesting questions, and wait for someone to answer them.

Someone asked *Quora*, "What is it like to have an understanding of very advanced mathematics? I'm interested to hear what very talented mathematicians and physicists have to say about 'what it's like' to have an internalized sense of very advanced mathematical concepts, just what it really feels like, to be a professional mathematician".

That question is one that I myself have been trying to answer, for years. The response by Anonymous is by far the best thing on this question that I have ever seen. I could not squeeze these excerpts into the text of this article, so here they are, as an Appendix.

Answer by "Anonymous":

> As you get more mathematically advanced, the examples you consider easy are actually complex insights built up from many easier examples.
>
> Once you know these threads between different parts of the universe, you can use them like wormholes to extricate yourself from a place where you would otherwise be stuck.
>
> The accomplishment a mathematician seeks is finding a new dictionary or wormhole between different parts of the conceptual universe.
>
> You can answer many seemingly difficult questions quickly. But you are not very impressed by what can look like magic, because you know the trick. The trick is that your brain can quickly decide if a question is answerable by one of a few

---

[9] http://www.quora.com/Mathematics/What-is-it-like-to-have-an-understanding-of-very-advanced-mathematics

powerful general purpose 'machines' (e.g., continuity arguments, the correspondences between geometric and algebraic objects, linear algebra, ways to reduce the infinite to the finite through various forms of compactness) combined with specific facts you have learned about your area. The number of fundamental ideas and techniques that people use to solve problems is, perhaps surprisingly, pretty small.

You are often confident that something is true long before you have an airtight proof for it (this happens especially often in geometry). The main reason is that you have a large catalogue of connections between concepts, and you can quickly intuit that if X were to be false, that would create tensions with other things you know to be true, so you are inclined to believe X is probably true to maintain the harmony of the conceptual space. It's not so much that you can imagine the situation perfectly, but you can quickly imagine many other things that are logically connected to it.

You are comfortable with feeling like you have no deep understanding of the problem you are studying. Indeed, when you do have a deep understanding, you have solved the problem and it is time to do something else. This makes the total time you spend in life reveling in your mastery of something quite brief. One of the main skills of research scientists of any type is knowing how to work comfortably and productively in a state of confusion.

Your intuitive thinking about a problem is productive and usefully structured, wasting little time on being aimlessly puzzled. For example, when answering a question about a high-dimensional space (e.g., whether a certain kind of rotation of a five-dimensional object has a "fixed point" which does not move during the rotation), you do not spend much time straining to visualize those things that do not have obvious analogues in two and three dimensions. (Violating this principle is a huge source of frustration for beginning maths students who don't know that they shouldn't be straining to visualize things for which they don't seem to have the visualizing machinery.)

When trying to understand a new thing, you automatically focus on very simple examples that are easy to think about, and then you leverage intuition about the examples into more impressive insights. For example, you might imagine two-

and three-dimensional rotations that are analogous to the one you really care about, and think about whether they clearly do or don't have the desired property. Then you think about what was important to the examples and try to distill those ideas into symbols. Often, you see that the key idea in the symbolic manipulations doesn't depend on anything about two or three dimensions, and you know how to answer your hard question.

As you get more mathematically advanced, the examples you consider easy are actually complex insights built up from many easier examples; the "simple case" you think about now took you two years to become comfortable with. But at any given stage, you do not strive to obtain a magical illumination about something intractable; you work to reduce it to the things that feel friendly.

To me, the biggest misconception that non-mathematicians have about how mathematicians think is that there is some mysterious mental faculty that is used to crack a problem all at once. In reality, one can ever think only a few moves ahead, trying out possible attacks from one's arsenal on simple examples relating to the problem, trying to establish partial results, or looking to make analogies with other ideas one understands. This is the same way that one solves problems in one's first real maths courses in university and in competitions. What happens as you get more advanced is simply that the arsenal grows larger, the thinking gets somewhat faster due to practice, and you have more examples to try, perhaps making better guesses about what is likely to yield progress. Sometimes, during this process, a sudden insight comes, but it would not be possible without the painstaking groundwork.

You go up in abstraction, "higher and higher". The main object of study yesterday becomes just an example or a tiny part of what you are considering today. For example, in calculus classes you think about functions or curves. In functional analysis or algebraic geometry, you think of spaces whose points are functions or curves—that is, you "zoom out" so that every function is just a point in a space, surrounded by many other "nearby" functions. Using this kind of zooming out technique, you can say very complex things in short sentences—things that, if unpacked and said at the zoomed-in level, would take up pages. Abstracting and compressing in this way allows you to consider extremely complicated issues while using your limited memory and processing power.

Learning the domain-specific elements of a different field can still be hard—for instance, physical intuition and economic intuition seem to rely on tricks of the brain that are not learned through mathematical training alone. But the quantitative and logical techniques you sharpen as a mathematician allow you to take many shortcuts that make learning other fields easier, as long as you are willing to be humble and modify those mathematical habits that are not useful in the new field.

You move easily between multiple seemingly very different ways of representing a problem. For example, most problems and concepts have more algebraic representations (closer in spirit to an algorithm) and more geometric ones (closer in spirit to a picture). You go back and forth between them naturally, using whichever one is more helpful at the moment.

Indeed, some of the most powerful ideas in mathematics (e.g., duality, Galois theory, algebraic geometry) provide "dictionaries" for moving between "worlds" in ways that, ex ante, are very surprising. For example, Galois theory allows us to use our understanding of symmetries of shapes (e.g., rigid motions of an octagon) to understand why you can solve any fourth-degree polynomial equation in closed form, but not any fifth-degree polynomial equation. Once you know these threads between different parts of the universe, you can use them like wormholes to extricate yourself from a place where you would otherwise be stuck.

Understanding something abstract or proving that something is true becomes a task a lot like building something. You think: "First I will lay this foundation, then I will build this framework using these familiar pieces, but leave the walls to fill in later, then I will test the beams ..." All these steps have mathematical analogues, and structuring things in a modular way allows you to spend several days thinking about something you do not understand without feeling lost or frustrated. (I should say, "without feeling unbearably lost and frustrated"; some amount of these feelings is inevitable, but the key is to reduce them to a tolerable degree.)

In listening to a seminar or while reading a paper, you don't get stuck as much as you used to in youth because you are good at modularizing a conceptual space, taking certain calculations or arguments you don't understand as "black boxes", and considering their implications anyway. You can sometimes make statements you know are true and have good intuition for, without understanding all the details. You can often detect where the delicate or interesting part of something is based on only a very high-level explanation. You are good at generating your own definitions and your own questions in thinking about some new kind of abstraction.

On the other hand, you are very comfortable with intentional imprecision or "hand-waving" in areas you know, because you know how to fill in the details.

[After learning to think rigorously, comes the] "post-rigorous" stage, in which one has grown comfortable with all the rigorous foundations of one's chosen field, and is now ready to revisit and refine one's pre-rigorous intuition on the subject, but this time with the intuition solidly buttressed by rigorous theory. (For instance, in this stage one would be able to quickly and accurately perform computations in vector calculus by using analogies with scalar calculus, or informal and semi-rigorous use of infinitesimals, big-O notation, and so forth, and be able to convert all such calculations into a rigorous argument whenever required.) The emphasis is now on applications, intuition, and the "big picture". This stage usually occupies the late graduate years and beyond.

In particular, an idea that took hours to understand correctly the first time ("for any arbitrarily small epsilon I can find a small delta so that this statement is true") becomes such a basic element of your later thinking that you don't give it conscious thought.

(From QUORA, Thursday, Dec 15, 2011)

References

Alperin, R. C. (1987). "Heron's area formula". In: *The College Mathematics Journal* 18, pp. 137–138.

Azzouni, J. (2005). "Is there still a Sense in which Mathematics can have Foundations?" In: *Essays on the Foundations of Mathematics and Logic*. Ed. by G. Sica. Milano: Polimetrica, pp. 9–47.

Behrens, C. (2012). "Empiricism: An Environment for Humanist Mathematics". In: *Journal of Humanistic Mathematics* 2.1.

Byers, W. (2007). *How mathematicians think*. Princeton: Princeton University Press.

Changeux, J.-P. and A. Connes (1998). *Conversations on Mind, Matter, and Mathematics*. Ed. by M. B. DeBevoise. Princeton: Princeton University Press.

Connes, A. (2008). "Advice to a Young Mathematician part III". In: *Princeton Companion to Mathematics*. Ed. by T. Gowers. Princeton: Princeton University Press, pp. 1005–1007.

Connes, A., A. Lichnerowicz, and M. Schützenberger (2001). *Triangle of thoughts*. American Mathematical Society.

Dehaene, S. and E. M. Brannon, eds. (2011). *Space, time and number in the brain*. San Diego, CA: Academic Press.

Dunham, W. (1990). *Journey through Genius*. New York: Penguin Books.

Gonthier, G. (2008). "Formal proof—The Four Color Theorem". In: *Notices of the American Mathematical Society* 55.11, pp. 1392–1393.

Hales, T. (2008). "Formal proof". In: *Notices of the American Mathematical Society* 55.11, pp. 1370–1380.

Hardy, G. H. (1929). "Mathematical proof". In: *Mind* 38.149, pp. 1–25.

Harris, M. (2008). ""Why mathematics?" You might ask". In: *Princeton Companion to Mathematics*. Ed. by T. Gowers. Princeton: Princeton University Press, pp. 966–977.

Heintz, B. (2000). *Die Innenwelt der Mathematik*. Wien: Springer.

Hersh, R. (1997). *What is mathematics, really?* Oxford: Oxford University Press.

— (2007). "On the interdisciplinary study of mathematical practice, with a real live case study". In: *Perspectives on Mathematical Practices. Bringing Together Philosophy of Mathematics, Sociology of Mathematics, and Mathematics Education*. Ed. by B. Van Kerkhove and J. P. Van Bendegem. Vol. 5. Logic, Epistemology, and the Unity of Science. Dordrecht: Springer. Chap. 13.

— (2011). "Mathematical Intuition". In: *Logic and Knowledge*. Ed. by C. Cellucci, E. Grosholz, and E. Ippoliti. Cambridge Scholars.

Johnson-Laird, P. N. (1983). *Mental Models: Towards a Cognitive Science of Language, Inference, and Consciousness*. Cambridge, MA: Harvard University Press.

Keynes, J. M. (1971). "Newton, the Man". In: *The Collected Writings of John Maynard Keynes Vol. X*. Macmillan, St. Martin's Press and the Royal Economic Society, pp. 363–374.

Larvor, B. (2012). "How to think about informal proofs". In: *Synthese* 187.2, pp. 715–730.

Monk, R. (1991). *Ludwig Wittgenstein: The Duty of Genius*. London: Vintage.
Nieder, A. (2011). "The Neural Code for Number". In: *Space, time and number in the brain*. Ed. by S. Dehaene and E. M. Brannon. San Diego, CA: Academic Press, pp. 103–123.
Peirce, B. (1879). *Linear Associative Algebras*. Lithograph.
Petitot, J. (2005). "Review of Conversations on Mind, Matter, and Mathematics". In: *Mathematical Intelligencer* 27.4, p. 48.
Rav, Y. (2006). "Philosophical Problems of Mathematics in the Light of Evolutionary Epistemology". In: *18 Unconventional Essays on the Nature of Mathematics*. Ed. by R. Hersh. New York: Springer, pp. 71–96.
Ruelle, D. (2007). *The Mathematician's Brain*. Princeton: Princeton University Press.
Wedberg, A. (1955). *Plato's philosophy of mathematics*. Almqvist & Wiksell.

CHAPTER 4

Why Philosophy and the Humanities Matter a Lot

Rik Pinxten

1 The concept of the 'two cultures'

In 1959 C. P. Snow gave his famous lecture on 'The Two Cultures' (published a year later; Snow 1959). He stated that natural sciences and the humanities including philosophy, are growing more and more apart, to the extent that the practitioners of both are no longer acquainted in a relevant or substantial way with the concepts, theories or discussions of the other 'culture'. Some years later the book got translated in French (in 1968), which triggered a double paper by later Nobel laureate I. Prigogine (Prigogine and Stengers 1976, 1977). The latter argued that a lack of basic scientific knowledge in the humanities and the arts, and the shallowness of cultural education among natural scientists, is becoming a widespread feature of the time. They deplore this evolution, because they hold, like Snow, that both types of knowledgeability are not only valuable in themselves, but important for the quality of knowledge in general.

At the time of Prigogine and Stengers's articles, a more general depreciation of rationality was becoming fashionable, leading to what have later been called the 'science wars' and the 'culture wars'. The authors cite J. Monod (Monod 1970), who pointed to a real crisis situation once both 'cultures' get estranged from one another. All seem to agree that the history of science makes abundantly clear how important it is for scientists to have a good education in history, culture and the arts. They emphasize the contribution of each 'culture' to the well-being of the other, and to the development of good thinkers in general.

All this was written on the verge of the postmodernist period, which announced a head-on attack on the authority of science and rationality. Monod and Prigogine must have seen the postmodernist wave coming, and stressed that any lack of knowledge only entails stupidity and loss of quality. It is not freeing people from chains or Master discourses. Decades later, on the occasion of the 50th anniversary of Snow's book, Jean Paul Van Bendegem picks up again the idea of the two cultures (Van Bendegem 2009), and makes remarks which resonate those of Prigogine and Stengers in their papers of 1976–77. Van Bendegem makes commendable remarks on the issue and continues where Prigogine and Monod had left off. He starts from a somewhat different perspective, in that meanwhile philosophy of science and the social studies of science had launched debates which add significantly to Snow's views on things. Although Van Bendegem's arguments are pretty much in line with Snow's and Prigogine's views, his criticism on the present state of affairs goes a step beyond theirs. Indeed, things got worse since the 60's and 70's of the past century. For one thing, philosophers became more and more professionals in their own discipline, working on texts and opinions of fellow philosophers while being progressively incompetent in the scientific disciplines themselves: they live and think at some distance from scientific practice and many of them never did any research in those fields. They are thus exclusively reasoning in a meta-world. This triggered the biting critique of Sokal and Bricmont (1998). The physicist Sokal denounced the blatant incompetence of so-called scientists in the humanities by getting utter nonsense papers of himself published in esteemed journals of philosophy and the humanities. The so-called 'science wars' ensued. However, as Prigogine and Stengers remarked and as Van Bendegem keeps on arguing, the lack of knowledge applies to both cultures. Artists and scholars in the humanities cling to a 19th century view on natural sciences, if they have any knowledge of them at all, but scientists have an equally uneducated and outdated frame of reference on the former.

One effect of this growing gap between the 'two cultures' is that there is a widespread blindness about the situatedness of scientific knowledge in society. Put simply, science is not a species or a natural phenomenon, but rather a human activity and product. Like all human activities and products, qua knowledge it is 'situated' (Lave and Wenger 1991), and hence has roots in a cultural and social tradition. The founding fathers of modern science may have lived under the illusion that scientific knowledge was absolute, pure and context-free, but today we are well informed about the fallibility, the at best relative certainty and the contextuality of the scientific enterprise and of its products. The impact of these constraints on the status of the knowledge produced should hence be studied, from a less naive perspective than the founding fathers did. In my opinion, many earlier scholars of both 'cultures' were not able to break loose from the theological and/or religious views on knowledge in their generation, and hence replaced one orthodoxy by another one. The second orthodoxy (the scientific view) may well yield better founded and more dependable knowledge than the first one, but the attitude of orthodoxy is almost the same, that is equally blind and wrong. Debates about the 'real' truth of scientific knowledge against the 'false' lore of religion are still rampant, and opponents often seem to forget that their mindset has remained strikingly unchanged throughout these controversies. The classical positivist would continue to claim that science produces the 'truth', although philosophy of science and the social studies of science at the very least deeply question this 'received view' (see e.g. Suppe 1977). In the past decades T. Kuhn, P. Feyerabend and other philosophers, but also S. Restivo (see chapter 7 of this volume, eds.), K. Knorr, B. Latour, I. Stengers and many other social scientific scholars, contextualized the production of knowledge in time, in cultural and in social-historical complexes. If anything, knowledge is grounded, as far as we know (Campbell 1989). The more modest, and at the same time more relevant and powerful view on scientific knowledge, in my view, proves to be that of another Nobel laureate, namely R. Feynman. Summarizing his view: We have more or less dependable knowledge in science, which remains always up for correction and falsification. Probably the only thing we know for sure is that we have no absolutely certain knowledge at all (Feynman 1985).

At this point, I pick up Van Bendegem's remarks, and flesh out three suggestions made in his book (Van Bendegem 2009):

- Yes to fallibility, no to relativism.

- Argumentation is important for all scientific work.

- Science and society (culture) should be linked, also in education.

2 Fallibility, not relativism

Fallibilism is one of these attitudes which will oppose believers and most non-believers, at least in the context of the religions of the book. Something like a culture or a cultivation of doubt, let alone a high priority status for doubt and criticism is abhorred in the religions of the book. Yes, discussions over the range of meanings of a term, or eventually over the order of the books which are said to be transferred over the generations, can often (but not always) be granted a certain place in these religions. But there is a limit to doubt and criticism, and the dogmas and received views of each religion will often make that limit explicit. Mystics who disregard such limit will be persecuted, as well as many a scholar who dares to interpret the holy texts in too liberal a way. Theologians will be condemned and forbidden to address audiences for a certain period (Boff, Schillebeeckx, Kung are only the more famous examples of a very long list in Christianity). Believers who take the risk to stretch the limits which the institutes have defined will be called heretics, and more often than not risk their very lives. I referred to this feature in the religions of the book as their 'intrinsic fundamentalism' (Pinxten 2010). It is, I believe, an important task for the institutes in these religions, to scrupulously check on this quality: the tradition will fossilize and suffocate religiosity when it allows the intrinsic fundamentalism to overrule any margin of interpretation and hence prevent changes to be coped with. History has proven that this task of safeguarding a certain degree of freedom or openness is a difficult but necessary one.

In their ambition to represent an undoubtable truth, these religions have been transferring the principle of absolute truth and their suspicions about doubt and fallibility to all knowledge. In the usual discourse this is referred to as the universal status of theories and concepts which originate from the religious source of knowledge. The doom that was seen to threaten any critic or doubter is then captured by relativism. Insecurity, lack of any clear or absolute referent, and hence relativism which in that tradition equals not-knowing.

The opposition between universalism and relativism has thus been obvious and basically taken for granted for generations. It is only in recent research that this dichotomy has been fundamentally questioned. For instance, the study of color has offered an interesting case on this issue. Since more than a century psychophysiology has mapped very systematically color perception at the organic level of the eye: after Newton's work in optics we know that the spectrum of physical color can be subdivided endlessly, by differentiating ever more values on three dimensions. Depending on the qualities of the device of registration (the eye, a microscope, and so on) the spectrum can be divided in an infinite amount of subcategories of color. The optical theory of color is universal: it applies to all light we know of. The psychophysiologists then studied the qualities of the human eye, and came to the conclusion that the human eye can only see a part of the physical color spectrum. Moreover, the eye discriminates between color categories in a discontinuous way: it groups shades of a color together in a discrete group, which is distinct from the next one. The total spectrum of color is thus seen much in the way of the color chart one finds in a paint shop: the discrimination of over 400 color categories the eye can distinguish. Between each color patch and the adjacent ones no further in-between differentiations can be made, although we know from optics that in the light beam an infinite range of differentiations is actually there . At the organic level, a second, but different set of universal characteristics of color thus obtains: the 'color chart' is seen by all humans (except the color blind subjects) in the same way, due to the qualities of the eye as a receptor system. Today we use a standardized representation of this psychophysiological level of color discrimination in the so-called Munsell color card.

On a third level, however, relativism enters on the scene. In a large scale linguistic anthropological study, Berlin and Kay (1969) used the Munsell card to investigate in what ways color categories are linguistically and culturally recognized and used in some 166 languages and cultures around the globe. Their aim was to determine in what ways categorically different colors are dealt with by 'the mind's eye', i.e., in languages and cultures. They called these the 'basic color categories'. It proved to be the case that one obtains a range of basic color categories, starting with only two and allowing for at the most eleven such basic categories or color terms. Beyond this small number, one finds only combinations: blueish green, Bordeaux red, and the like. But some languages and cultures would only have two basic terms/categories, dividing the whole Munsell card in two halves. The left half of the card (from blue to green and brown) would be covered by one term, and the right half (from yellow over orange to red) by the other term. In languages and cultures where the maximum of eleven basic color terms would be used, the whole card is divided in eleven islands of groups of color patches on the Munsell card. Over all, languages and cultures showed a uniform division of the Munsell card. Color research is intriguing because the humanities have been doing it for more than a century and the amount of data and of theoretical processing of the latter is enormous. In other domains, such as kinship studies (Goodenenough 1970), plant taxonomies (Conklin 1971) and spatial notions (Pinxten, van Dooren, and Harvey 1983) the same 'double' level of knowledge was found. That is to say, the universal features at the physical and/or the physiological level is paralleled by culture specific categorization.

The important thing to be learned from such studies, I claim, is that universalism and relativism are not mutually exclusive orthodox positions, but rather complementary ones. That is to say, universalism seems to obtain at the level of physical and (differently) physiological reality, but not necessarily at that of linguistic or cultural issues. Propositions at all these levels may well be understood as complementary with each other, because they speak about different aspects of reality. It is only with a blind (and dogmatic) a priori of reductionism of the sociocultural to the physical (or biological) that universalism is clearly and always exclusive of relativism. Or, in other words, it is only within this a priori and hence possibly unwarranted reductionist frame that the absoluteness of both a relativistic and a universalistic stance becomes obvious. Van Bendegem attacks reductionism at very considerable length in the 2009 book.

3 Argumentation

C. Perelman, together with S. Toulmin, was one of the major figures in the new line of argumentation theory or rhetorics, which picked up the age old tradition after the Second World War. In his path breaking book *Traité de l'Argumentation. La nouvelle rhétorique* (Perelman and Olbrechts-Tyteca 1957), he attempts to give rhetorics more solid foundations than those the Ancients used to work with, drawing on the humanities and the social sciences to found a new theory of argumentation. At the same time he makes a strong plea for the adoption of argumentation as a constitutive part of all scientific work. He even makes the point that only theology works by means of convictions (with a resolute absoluteness attached to them) from which inferences can be deduced in a rather straightforward logical line.

Even mathematics, Perelman remarks, is likely to be closer to the empirical sciences and hence in need of argumentation through persuasion, than aprioristic theology. But most certainly, in his view, the sciences have a degree of persuasive argumentation in their procedures, similar to and at the same time different from politics, ethics or aesthetics. In combination with the former paragraph, it is obvious that apriorism, also in the realm of logic and mathematics and surely in that of all empirical sciences, is a debatable philosophical stance for modernists, including scientists. It does not come as a surprise that Van Bendegem (2009) makes the point that the elimination of argumentation styles and theories from the standard curriculum of the education of a researcher (both in the natural sciences and in the humanities) is certainly to be deplored.

In line with Perelman I would argue that this is an impoverishment in the skills the researcher will need for his or her job. Van Bendegem (2009, p. 126) makes the point that the loss is even more substantial: the researcher is thus estranged from his or her more varied life as a citizen, family person or societal being, since the skills of argumentation will be needed in 'real' life, also by the researcher. On top of that, of course, the lack of training in argumentation theories is a loss since scientific work, as argued above, is situated in this larger societal context. The researcher who is unable to plead his or her cause there, is becoming more and more a blind and dumb specialist, who is at the mercy of other 'speakers' for the contextualization of research, its worth, its relevance and its beauty. In the context of a growing impact of multinational corporations on choices made for and in the name of science, this is a dangerous loss.

Some events illustrate this point today: our colleague P. Verhaeghe abundantly describes how pharmaceutical mega-industries dictate what should be on the agenda of health studies, and the worldwide development of bound-science in Recombined DNA yields secrecy and lack of debate among scientists together with a brutal standardization of agribusiness plantations throughout the world by some of the multinationals in this field (Verhaeghe 2012). Confronted with a small and rather amateurish protest against such practices in genetically manipulated potatoes research in Belgium, the researchers were given a speed course in argumentation by the corporations in order to be able to make a more professional impression in debates in the media. Two decades ago, similar courses were obligatory for researchers working in the nuclear branch in order to counter arguments in public events. Typically, the courses are elementary and teach the poor scientists how to play the part without going under. This is different, of course, from being able to reason on choices one makes, to weigh arguments and see and discuss the broader context, and so on. It is rather a speed course on not turning into a sitting duck when spotted by an angry hunter who feels bedeviled by the big bad wolf who got all the opportunities. One need not be a hardboiled anarchist to say that this sort of rescue procedures for the benefit of 'science' are a far cry from free, democratic scientific research for the benefit of all.

4 The link between science and society

The final chapters of Van Bendegem's book on the Two Cultures, are a heartfelt plea for the reintegration of scientific research in society. And hence, it is a plea to the benefit of the humanities and the social sciences, because they are the best we have to get more insightful knowledge of these highly complex phenomena we call society, religion culture, and what have you. At this point, I agree with him wholeheartedly, but at the same time I tend to think my friend and colleague is much too friendly with the foes that are around. One such demon I want to stress here is that of managerial craze (as Verhaeghe 2012 calls it repeatedly). I end this paper with some comments, based on my own experiences in the course of a career. My justification for doing this is that, if I do not speak up as an older scholar who has nothing to loose, the demon will spread its ugly wings ever wider and suffocate creativity and intelligence still more. I want to argue by means of a witness report, with myself in the role of the witness. This is an illustration of Perelman's perspective on argumentation theory. I have to break out of the circle of rules and agreements of the present managerial definition of university research to make the argument, thus following a path that is rather similar to that of the first modern scientists who had to pierce holes in the walls of the theological and clerical context they were supposed to respect as the limits of their scope on reality. (Verhaeghe 2012)

When I came to the university as a student, I felt at home. Not only did I have the impression that this was a context where I could flourish, but also that genuine discussions and questions might be treated and allowed here in an atmosphere of respect and genuine interest. That is to say, I felt there was room for research and questioning, for developing what I later called 'a culture of critique'. Throughout my life I held the conviction that creativity in thought, in artistic expression and in interpersonal relationships is what makes human dignity real. Also, each generation adds and interprets this margin of creativity.

At the same time I felt then as a student and now as a retiring professor that a blind academic treatment of important issues was dangerous and alienated. I mean, I immediately experienced that there was a good chance that scholars and such would hide from the real world in their safe bastion, with the normative and obviously political message that they were dealing with 'the truth', after all. Funny pretension, of course, but it existed and still does. At the time it was believed that science would yield real and true knowledge, as if detached from anything human, including from emotions and context.

In the 1960s and later on, this pretension was heavily attacked and subsided to a certain degree. After the collapse of the Soviet Union in 1990 however, the new antisocial religion of exclusive managerial thinking stood up and installed a somewhat new, but equally pretentious and unfounded form of blind positivism, horribly nicely caught in the brutal slogan of 'measuring is knowing'. In other words, instead of 'do not think but obey the authority of predecessors' of the 1950s we now live under the slogan 'do not think but obey the numbers which others tell you to be relevant'. The interested reader is referred to the appalling book edited by Chomsky on these issues (Chomsky 1996).

Over the years I was allowed to continue my search because there were a few enlightened and indeed rather courageous professors around. Since I studied moral science and philosophy in Ghent, I refer to great thinkers at that time like Leo Apostel and Jaap Kruithof. Also Rudolf Boehm. The latter was not well understood by me, but he kept on intriguing me: this is a strong personality, who fills in the culture of criticism in a particular way, disregarding the petty bourgeois gossip on his person and his thoughts in a rather aristocratic way. Relentlessly he kept going on throughout his life to write his philosophy, as a sort of anarchistic nobleman. He was sticking his neck out on the side of the students at the time of the student revolts, which almost cost him his job. However, in the recent recuperation of the '68–'69 movements by media and politicians his role was blatantly overlooked. *L'histoire se répète*, I thought. Moreover, Boehm was at the toplevel of continental philosophy in Germany, France and the USA, and in his classes we were fascinated but rather estranged by his teachings.

Leo Apostel and Jaap Kruithof were more easily connecting to me. They opened our minds and often drew us in discussions and questionings which led onto a world of scientific and political thinking that was mostly unknown in Flanders then. They had the broadness of mind to allow me and some others, like Diderik Batens (see chapter 2 of this volume, eds.) and Marc de Mey, and later also Jean Paul Van Bendegem and Werner Callebaut, to do their thing and swim in the intellectual waters that were out there. What a tremendous chance I had to meet these people rather than the usual grey scholars of Ghent university of that time. Without doubt these few people allowed me to follow my inclinations and hence develop a scientific perspective that was not theirs, linked with a much broader horizon than Flanders then or now. The culture of criticism, which is so dear to me, indeed needs a global scope by now. Having said that, I should add that it might be threatened today on a global scale as well, by the new religion of the managerial trend, uncritically at the service of what Abram de Swaan (Amsterdam) has called the ideological period of "marketism" we have now landed in (after those of theocentrism and marxism). Neoliberal dehumanisation, I called it once. Anyway, Leo and Jaap allowed me to carve my road (and Etienne Vermeersch gave me a roof over my head) although they remained very skeptical about the path I was taking.

Being the positive and optimistic person I am, I went for it. Instead of focusing on a particular subquestion of a presumably great thinker in order to add a footnote or two to his master's voice, I formulated a wild idea which has been with me for almost my entire career: what if other cultures, religions or languages would think, speak and act in a world which differs importantly from ours? How come they were able to survive, when they were believed to be 'wrong' by centuries of western philosophers and theologians? How can I approach such questions and hence promote some sort of pluralism in this western dominated and colonial world? This was and is an epistemological question for me, with political ramifications to be sure.

Within this broad scope I went up to and spoke and collaborated with great scholars, with international reputations, often residing at very prestigious universities, over the years. Not out of vanity, but just to honour them by naming a few: there were Claude Lévi-Strauss, Mohammed Arkoun and Pierre Bourdieu in France, Joseph Needham in the UK, academics like Don Campbell, Laura Nader, Clifford Geertz, Robert Rubinstein, Oswald Werner, Jim Wertsch, Richard Adams, Sal Restivo (see chapter 7 of this volume, eds.), Nelson Goodman and Johannes Fabian, but also novelists like Richard Powers and Rhoda Lerman, or mathematicians like Reuben Hersh (see chapter 3 of this volume, eds.), Marylin Frankenstein and many others in the USA, as well as many people in other parts of the world. Some of them became friends like Don Campbell, Sal Restivo, Laura Nader and Arie de Ruijter. But what strikes me most of all is that these scholars showed interest in my unusual research question and gave me time to write them, visit them, talk with them and exchange texts and insights for years on end. I mention no colleagues in Flanders, because naming some and forgetting others here hurts. But they are there, of course. Also, and this is very important for me as a person and as an academic: the young people were and are there.

Whatever their busy schedule, they actually reserved time for me, a scholar from a small and peripheral place like Flanders, Belgium. They were not doing this out of charity, I believe, but because they recognized a searching soul or wrestled with a similar question. Because I experienced each and all of them as persons who were driven by a genuine motivation to investigate and acquire some knowledge. Not by status, power or another extrinsic motivation: they were still the boys and girls with deep questions, whatever their status and position. And hence they first allowed time and space for me who went up to them with my question, and often continued to do so along this road. I was an optimist and I remained that way because of these experiences, although the pettiness and shortsightedness of many others along the way sometimes had me almost detracted from my search. However, my conviction remains that science without an intellectual scope is at best sophisticated arty-farty science, at worst a waste of human effort.

After two decades of ripening in scientific work, with an atmosphere in the 1970s and 1980s that allowed for many initiatives, the 1990s came around. I had been doing my thing with colleagues and expanding in several directions. I had the joy to organize international symposia with great people some of whom I mentioned earlier, but also with Ilya Prigogine, André Gingrich or Ulf Hannerz. But with the collapse of the former Soviet empire things started to change: The West became righteous again in a sort of drunken zeal of victoriousness. And of a sudden, the intellectuals were not needed anymore: The enemy had been defeated, hence critical thinking should become less prominent while business should get the supreme role. In order to try and optimize this ideology for the emerging corporations and multinational banking firms, the religion of managerial governance was substituting for the intellectual traditions that were allowed during the cold war. Nobody wrote better about all this than Noam Chomsky. What he called 'the new Mandarins' took over and installed a system of control, without ever allowing to develop a vision or a long term view on science and society. Looking back on these past two decades, what Sennett (2008) calls 'the Bismarck model' was installed in Europe. Everything nowadays is measured and controlled, but research —let alone public debate— about the grand scheme or about the humanistic value of what we are doing went largely out the window. Even more, it was gradually said to be nostalgic or outdated to busy oneself with quality of life, with grand theories, or with updating humanism, given the interdependent world we currently live in. In the political arena, we were led to live with fear since 2001, and with the stupidity and blindness of a reemerging neo-nationalism in the West. Of course, since the 1990s the West has rapidly been loosing its superior position concerning economic and political power in the world. But instead of engaging the intellectuals in a search for a new and globally negotiated cosmopolitanism (which even an intelligent Republican like Henry Kissinger called for), we are becoming more and more insulary in our mentality. Why do we not use the facilities we have, in casu our splendid educational system, to engage the younger generation in an openminded and broad perspective upon our place in a pluriform and deeply interdependent world. Rather, we see tendencies to hide from the world in what Bourdieu rightly called the 'scholastic view': We see more and more control on each other's formal and marketable behaviour and its products, with a nearly total abandonment of discussion and thought about the goals and the meaning of life in the whole enterprise. Very much like the scholastics of the Middle Ages, who engaged in the control of the level of mindless belief by the followers and by themselves, rather than in open discussion, let alone innovating projects in the christian heritage they were standing for. I think this is a serious threat to thinking, and to the quality of life in the West. Its translation in the us-them righteous political reduction we witness today is allowed room for flourishing because the higher education and research centres of (in our case) Flanders are abdicating from the tradition of intellectual critique, in my view. These sorts of messages have triggered some reactions already. One is that the arts will save the world. Another is that the market will make everything alright in balancing out what is good for all.

The first reaction is not popular with artists. They are flagellated even more by the specter of managerial marketism than other domains of our society. My good friends in the arts tell me that things are not looking up these days. Nevertheless, that domain of human creativity which we call art has been tremendously important in the history of the West, for society, for religion and for the sciences. The second reaction is too stupid for words; poverty increased more rapidly the past two decades of marketism than ever before, with the 2008 bank crisis causing over 40 millions of hunger deaths in the past 3 years in the south, according to Nobel Prize winner Stiglitz (Stiglitz 2010). Education, health and old people's care in the UK and in the USA are in rapidly growing decay over the same period. And so on. My gut-reaction to this trend was that I began writing books in Dutch, apart from the more academic stuff. Today I am not sure this will help to safeguard or create an open society, but I am convinced that we should keep on doing this. Or make artworks, or be creative in other ways.

I mention all this because I am an optimist. In the past the culture of criticism managed to overthrow righteousness and dictatorships. We developed the arts, the sciences and some interesting social political thinking instead. Of course, nothing was or is eternal or absolute, but at the very least the culture of criticism secured that a continuous quality testing remained in place. I am an optimist, because I witnessed during my professional life that the promotion of creativity and humanist knowledge and culture is possible, provided the openness of mind and the recognition of valuable pluralism are safeguarded. I hope that the present period of self-righteous cocooning will soon be over and a new humanistic balance will be emerging. The challenges are there, and they are considerable. The young minds are here, but they hardly get a chance. It would be such a pity to throw away or not use the qualities that we have. The next generation will make the difference, and I sincerely hope that the broad road towards a panhumanistic and equal opportunity society will be on top of their agenda.

My conviction is that the humanities and philosophy matter if we want a humanistic and democratic future, with a relevant place for scientific research. Whether we will move in that direction is, however, an open question. I believe we will, once we manage to demolish the false religion of management. That is to say, if and when we start discussing views and perspectives in a deep way first, in a sort of Republic of the minds, engaging lay people and scientists alike. Afterwards, we can then apply some management tools to implement our choices, in an open way allowing for rethinking and change of choices at regular intervals. It runs in the tracks of the last pages of Van Bendegem's 2009 book in stating this. However, my political question is more straightforward: shall we take this turn? The answer is yours.

References

Berlin, B. and P. Kay (1969). *Basic Color Terms.* Berkeley: University of California Press.

Campbell, D. (1989). *Methodology and Epistemology for Social Sciences.* Chicago: University of Chicago Press.

Chomsky, N. (1996). *The Humanities in the USA during the Cold War.* New York: Simon and Schuster.

Conklin, H. C. (1971). *Classification.* New Haven: Yale University Press.

Feynman, R. (1985). *Surely You're Joking, Mr. Feynman.* New York: Simon and Schuster.

Goodenenough, W. (1970). *Description and Comparison in Cultural Anthropology.* Chicago: Aldine.

Lave, J. and E. Wenger (1991). *Situated learning: Legitimate peripheral participation.* Cambridge: Cambridge University Press.

Monod, J. (1970). *Le Hasard et la Nécessité.* Paris: Gallimard.

Perelman, C. and S. Olbrechts-Tyteca (1957). *Traité de l'Argumentation. La nouvelle rhétorique.* Paris: PUF.

Pinxten, R. (2010). *The Creation of God.* Frankfurt: P. Lang Verlag.

Pinxten, R., I. van Dooren, and F. Harvey (1983). *Anthropology of Space.* Philadelphia: University of Pennsylvania Press.

Prigogine, I. and I. Stengers (1976). "Les Deux Cultures Aujourd'hui, 1". In: *Nouvelle revue française* 316, pp. 42–54.
— (1977). "Les Deux Cultures Aujourd'hui, 2". In: *Nouvelle revue française* 317, pp. 41–48.
Sennett, R. (2008). *The Culture of New Capitalism*. London: ZED.
Snow, C. P. (1959). *The Two Cultures. Rede Lecture.* Cambridge University Press.
Sokal, A. and J. Bricmont (1998). *Fashionable Nonsense*. New York: Picador.
Stiglitz, J. (2010). *In Free Fall.* New York: Simon and Schuster.
Suppe, P. (1977). *The structure of scientific theories.* Chicago: University of Illinois Press.
Van Bendegem, J. P. (2009). *Hamlet en Entropie.* Brussel: VUB Press.
Verhaeghe, P. (2012). *Identiteit.* Amsterdam: De Bezige Bij.

CHAPTER 5

Mathematical Pluralism*

Graham Priest

Preface

It is a great pleasure to dedicate this essay to Jean Paul Van Bendegem on the occasion of his 60th birthday. I met Jean Paul on my first visit to Gent many years ago, and have seen much of him since, on other visits to Belgium, and on his one visit to Australia. This paper seems a highly appropriate one, given the plurality of Jean Paul's own many interests. Whatever his topic, I have always found his thoughts interesting and insightful (and his good humour refreshing). It was his work on finitism, indeed, that alerted me to the fact that inconsistent models of arithmetic may be more than just technical curiosities. I am, of course, pleased to see that the natural numbers stretch at least as far as 60, and I wish Jean Paul every success in continuing to prove that they stretch a great deal further.

1 The Variety of Mathematics

It is clear from a very cursory review of mathematical practice, past and present, that mathematicians concern themselves with a wide variety

---

*The material after the preface is reprinted from *Logic Journal of the IGPL* 21(1), pp. 4–13, 2013, by kind permission of Oxford University Press, on behalf of the International Interest Group in Pure and Applied Logics.

of investigations. They investigate the structure of groups, of random variables, of the complex plane, the natural numbers, infinite cardinals, and so on. There is, then, a pluralism of mathematical practices.

A natural thought—I presume a currently orthodox one—is that the pluralism is, in a certain sense, a superficial one. There is a single overarching mathematical theory, say Zermelo Fraenkel set theory, maybe with the Axiom of Choice. The various structures that are investigated are defined within this. The different practices are therefore all investigations of the structure of the set-theoretic universe delivered by ZF(C), or of various parts thereof.

The thought, though, does not survive long. First, mathematicians do not consider just what can be done within ZF(C). They consider extensions of ZF(C)—for example, with various large cardinal axioms. Next, it is not the case that all of standard mathematics fits into ZF(C) anyway. The programme of reducing mathematics circa 1900 to ZF(C) was spectacularly successful, but there are problems with later parts of mathematics. Category theory is an obvious example. Category theorists investigate the category of all sets, and even the category of all categories. For well known reasons, these do not live in the set-theoretic universe.[1] Of course, mathematicians realise this, and have suggested ways of accommodating these "overlarge" categories: we introduce proper classes (and proper classes of proper classes, etc.); or we invoke some large cardinal axiom, and then talk about the category of sets of some bounded rank. In the end, though, these devices just defer the problem; we have simply changed the subject. We still cannot apply category theory to *all* collections (by whatever name we choose to call them).

2 Non-Classical Mathematics

The failure of mathematics to fit into ZF(C) holds also for reasons much more radical. Let us call the sorts of mathematical investigations we have talked about so far, *classical mathematics*. How to characterise these is somewhat moot, but let us just take it that they may be pursued using classical logic.[2] There are also non-classical mathematics, where the underlying logics of the investigations are non-classical.[3]

---

[1] See, e.g., Priest (2006, ch. 2).

[2] Whether the underlying logic *must* be seen as classical is a different, and more contentious, matter.

[3] A referee objected that using the term 'logic' here is out of place, since logic has to do with truth-preservation, and these "logics" can't all encode truth-preservation. In fact, I think that they do preserve truth in those worlds which realise the characterisation of the objects of the investigation. More of this later. However, if one does

Several of these have appeared since they hey-day of mathematical reductionism. The most obvious example is intuitionist mathematics and its various branches: the theory of species, the intuitionist reals, the theory of the creative subject.[4] Clearly, there is no hope of reducing such investigations to a theory based on classical logic by giving explicit definitions of the intuitionist notions in classical terms. One might hope to be able to *interpret* all these theories by finding classical models. Thus, for example, one can interpret intuitionist logic in classical logic plus an *S4* modal operator.[5] But it appears to be impossible to interpret many intuitionistic theories in such a way. Thus, for example, some intuitionist theories of the reals contain both Brouwer's Continuity Theorem (every real-valued function defined over the closed unit interval [0,1] of a real variable is uniformly continuous on that interval) and Intuitionistic Church's Thesis (every total function from the natural numbers into the natural numbers is Turing computable). By suitable interpretations, one can understand each of these in classical terms; but not both together, by a construction of Specker.[6] Moreover, even if it were possible to interpret all intuitionist mathematics classically, this is *not* how intuitionistic mathematics is done. To view it in this way is therefore a *falsification* of the practice. It is as if one should claim that speakers of English are really speaking Latin, because everything they say can be translated into Latin.[7]

Matters are similar with respect to a more recent variety of non-classical mathematics: paraconsistent mathematics. There are now investigations of various inconsistent mathematical theories based on some paraconsistent logic or other: set theory based on naive comprehension, inconsistent arithmetics, inconsistent geometries (for example, of impossible pictures), to name but a few.[8] Again, there are various reductionist

---

not want to call this logic, I don't really mind. The question is only whether these are systematic investigations of something mathematically interesting using some notion of deduction.

[4]See, e.g., Dummett (1977).

[5]See Priest (2008, §6.10 (11) and §20.13 (11)).

[6]See Bridges and Richman (1987). Many thanks to David McCarty for helpful discussions on these matters.

[7]The plurality of mathematics is defended on the ground of constructive mathematics in Davies (2005); it is defended on the grounds of both constructive mathematics and category theory in Hellman and Bell (2006). A variety of mathematics, including classical and various constructive mathematics, is defended in Sambin (2011). Sambin does this, obtaining these mathematics by varying parameters within a "minimalist foundation" of mathematics (distinct from set theory). I do not see why, however, the variety of interesting mathematics must be constrained by *any* procrustian framework.

[8]E.g., Priest (2006, chs. 17, 18); Mortensen (1995, 2002); Weber (2010).

strategies that might sometimes come to mind. Thus, the inconsistent arithmetics are often defined by their (paraconsistent) models. So one can think of the investigation as one within standard model-theory. But this strategy does not apply to the development of inconsistent set theory, or of inconsistent geometries.

I interpolate that one does not have to be an intuitionist or a dialetheist to take intuitionist or paraconsistent mathematics to be legitimate. It suffices that these are interesting *mathematical* enterprises.

Non-classical mathematics are not only to be found in the developments of 20th century mathematics; they are also to be found in the history of mathematics before the 19th century "drive for rigour". An obvious example of this is the infinitesimal calculus in the 17th and 18th centuries. Quite self-consciously, mathematicians treated infinitesimals as non-zero at one stage of their proofs, and zero at another. Clearly, some paraconsistent reasoning strategy was being employed.[9]

Of course, the reasoning strategy was changed later with the invention of epsilon/delta methods by Cauchy and Weierstrass. Maybe this was a better way of doing things; maybe this is what earlier mathematicians were all really *trying* to do (though I doubt it). But the point remains: the proof-procedures were *changed*; mathematical practice was *revised*. The original practice was still, therefore, an example of a mathematical practice which was highly non-classical.

3 Mathematical Practices and Games

It would appear, then, there is an undeniable *pluralism* in mathematics. And the obvious question is how to make sense of this. In what follows, I will suggest a way.

Let us start by considering games. Pluralism is obviously true of games. Let us consider some salient facts:

1. There are many games; and any individual can play lots of different games. Thus, chess and draughts (checkers) are such games.

2. One of these might be more interesting than the other, more aesthetically pleasing, have a richer structure than the other. But, *qua* game, both are equally legitimate.

---

[9]The inconsistency of the early calculus has been contested by, e.g., Vickers (2007). His ground is essentially that there were few working in the area at the time who endorsed the existence of objects (infinitesimals, fluxations) with contradictory properties. Instead, there was disagreement, confusion, and uncertainty about the rationale for the procedures employed. This is hardly surprising, since the method employed *did* seem to depend on such an assumption. See Brown and Priest (2004), and Sweeney (2013).

3. Games have rules. The rules may have been made explicit, as in the case of chess and checkers; or they may only be implicit in a practice, as are many children's games (or as are the grammatical rules of a language).

4. The rules may be learned explicitly—as, normally, one learns chess; or they may simply be picked up by entering into the game and having one's actions corrected until, in the end, one just "has a feel" for what to do.

5. Whichever of these is the case, playing the game is just following the rules.

6. Typically, there is a point to following the rules: winning—which is not to say that a person must have the personal aim of winning to play the game; just that it's the institutional point.[10]

I want to suggest that mathematical pluralism is similar.[11] There is a plurality of mathematical practices: category theory, intuitionist analysis, inconsistent calculus. Each of these is governed by a set of rules—including inference rules—and engaging in the practice means following the rules. The (institutional) point of following the rules is establishing (proving) certain—hopefully interesting—things within the rules of the practice.[12] The rules may be explicit, as they typically are in contemporary mathematics; or implicit, as they were with number theory until the late 19th century. One may absorb the rules simply by being trained to follow them, as one learns a first language; arithmetic is usually learned in this way. Or one may learn the rules more reflectively, as one learns a second language; the way that a classically trained mathematician has to struggle with intuitionistic proof when they first meet intuitionist mathematics is like this. And just as with games, some practices may be more interesting, fruitful, or whatever; but all practices, *qua* practices, are equally legitimate.

One may balk at this point.[13] Not all practices are equally legitimate. In particular, some of the practices (the legitimate ones) serve to establish truths; the others do not. This raises the question of truth, to

---

[10]Of course, in some things we are inclined to call games there are no winners or losers. These, presumably, have other points.

[11]I stress that this is an *analogy*; I am not suggesting that to do mathematics is to play a game. There are obvious dissimilarities too.

[12]If one takes the things proved to be true of some domain of entities (as I will suggest below), one might think of the aim of the practice, alternatively, as establishing (interesting) truths about the objects in that domain.

[13]Indeed, one referee of a draft of this paper did.

which I will turn in due course. For the present, I note only that, as far as pure mathematics goes, mathematicians appear to be less interested in truth that truth-in-a-given-structure (or family of structures). Some structures are, of course, mathematically more interesting than others, have more natural applications, or wot not. But that is a different matter.

Something that speaks very strongly in favour of this view is the fact that it makes excellent sense of the phenomenology of mathematics. When one learns a game, such as chess, one is initially very conscious of the rules. ('This is a knight; now, how do they move?') Once one internalises the rules, one no longer thinks of them, however. They create a phenomenologically objective space, within which one just moves around. Similarly, when one learns a new mathematical practice, one has to concentrate very hard on the rules. ('This is a group; now, what properties does the group operator have?', 'We are doing intuitionistic mathematics; now, is this a legitimate inferential move?') But once the rules are internalised, the phenomenology changes, and we again find ourselves within an objective terrain within which we move around.

This seems an appropriate place to say something about formalism as a foundationalist view of mathematics. Perhaps the most plausible version of formalism is to the effect that mathematics is simply the development of formal systems; that is, mathematics is nothing more than symbol-manipulation in each such system.[14] This view is sometimes described by saying that mathematics is a game with symbols;[15] and so it might be thought that I am advocating a variety of formalism. Now, whilst there are certainly some similarities between this version of formalism and the view that I am suggesting here, there are crucial differences. For a start, there is no suggestion that an arbitrary formal system is a mathematical one: the system must have mathematical content. Nor is there a suggestion that every mathematical investigation is a formal system. People were "playing the game" of arithmetic for millenia before it was formalised. Perhaps it cannot even be formalised. (If it is essentially a second-order theory, it cannot.) Next, the view I am suggesting is quite compatible with the view that mathematical terms refer to objects of various kinds. (More of this in due course.) Finally, and to return matters phenomenological, the phenomenologies of doing mathematics and of manipulating symbols are quite different, as I have just stressed. When one learns a branch of mathematics initially, one may be doing little more that operating on symbols according to

---

[14]Something like this view is to be found in Carnap (1937) and Curry (1958). For a discussion of the various forms of formalism and their problems, see Weir (2011).

[15]See Horsten (2007, §2.2).

rules; but the phenomenology of a fully fledged mathematical practice is exactly one of acquaintance with the objects that the symbols (noun phrases) refer to.[16]

4 Mathematical Interactions

One might wonder what makes all these rule-governed activities *mathematics* once we have given up the hegemony of ZF(C), classical logic, etc. The obvious answer is that provided by Wittgenstein's *Investigations* for games themselves.[17] The plurality of mathematics are bound together by a family resemblance—and one, it might be added, whose bounds we are ever enriching and stretching.

There are tighter connections between the practices than mere family resemblance, however—at least sometimes. Albeit the case that there are different mathematical practices, some bits of mathematics "hang together". For example, we use arithmetic to count all kinds of mathematical objects, and we apply group theory in arithmetic and geometry. One might try to turn this observation into an objection against the view suggested here. Mathematical practices are essentially different from games because they can hang together in a way that games do not.

Of course, one could take the bits that hang together to be parts of one over-arching game (of ultimate truth), such as ZF(C), as, perhaps, some who work in the foundations of mathematics do. That most practicing mathematicians do so, I find hard to believe. (I suspect that most have no interest in the foundations of mathematics whatsoever, and would be hard pressed even to state the axioms of ZF(C).) In any case, the reduction of parts of mathematics to ZF(C) appears to be simply a *post facto* reconstruction of parts of mathematics. Actual history is, in fact, much more interesting; and the objection is simply oblivious to the many inter-relationships between games that there may be.

For a start, games can overlap. Thus, the rules of rugby union and rugby league have an overlap. The rules of one game can even subsume those of another. Thus, consider the game chess$^-$, which is the same as chess, except that there is no castling. The rules of chess subsume those of chess$^-$. So it can be with mathematical practices. Thus, as observed, in many parts of mathematics, objects are counted. For example, we may count the number of groups of a certain kind. This means that, in such practices, number theory, or at least an important part of it, is

---

[16] Hence, according to the view which I am putting forward, mathematics has a content. That formalism cannot account for the content of mathematics was the most substantial of Frege's arguments (in *Grundgesetze*) against the formalists of his day.

[17] See, e.g., §§ 65, 67.

subsumed.

Practices can also evolve from others practices. (We will note that the same is true of games in a moment.) Take group theory. In the 19th century, mathematicians such as Galois abstracted from the structures of numbers, geometries, and other things, to formulate the notion of a group. In writing down the axioms of group theory, and investigating their consequences, they initiated the practice of group theory. But this allows us to apply a theorem of group theory to any one of the areas from which it was abstracted. For the fact that certain moves can be made, generally, in the group-theoretic game means that they can be made, specifically, in the game for, say, number theory, since they are equally moves there. Group theory provides, as it were move-schemata.

Another example: one game can be absorbed into a larger game (which does not mean that the original game cannot be played in its own right). Thus, the usual 2-dimensional chess can be extended to a game of 3-dimensional chess—in such a way that the restriction of the 3-dimensional game any two-dimensional plane would coincide with the 2-dimensional game. In a similar way, one can see elementary number theory as being incorporated in analytic number theory.[18]

The chess example shows, by the way, that there may be natural and not so natural ways of extending a game. Thus the natural extension of ordinary chess into the third dimension still has, e.g., bishops moving along diagonals, rather than some other trajectory. Mathematics may be the same. Thus, assume (mootly) that to do arithmetic is to follow the rules of first-order Peano Arithmetic. The extension of this to second-order PA (or some axiomatisable fragment thereof), where axiom schemata are turned into quantified sentences, is clearly natural. Such an extension is much more natural than simply adding $\neg Con(PA)$ to PA.[19]

The history of mathematics provides a rich ground for those who would investigate the interaction between mathematical practices. But enough of this for now.

---

[18] This raises an interesting question about the identity of objects across practices. Suppose that we prove something about a number, n, using analytic number theory, which cannot be proved using elementary number theory. Are they the same n? I will turn to matters ontological at the end of the paper, but the answer is essentially yes. There is no reason why different practices must be about different objects. In the same way, different stories can be about the same object, e.g., Napoleon or Sherlock Holmes. In other words, the identity of an object need not be practice-bound.

[19] This should assuage the concerns voiced by Koellner (2011) concerning certain kinds of pluralism.

5 Further Objections

Let us move on to some other objections. According to this view, each of the multiplicity of mathematics is defined by a set of rules. One might worry that mathematics is not a rule-bound activity. Of course, in one sense it clearly is: proving is a tightly constrained activity. But, one might think, mathematicians break the rules sometimes. They may invoke new axioms or new methods of proof, not legitimised by the game being played. This, however, is easily accounted for. About 200 years ago, according to the standard story, during a game of soccer at Rugby School, someone picked up the ball and started to run with it. They broke the rules; but what they did was instigate a new game, rugby. In the same way, when someone breaks the rules of a mathematical game, and they are not simply making a mistake, they are, *ipso facto*, no longer playing that game. The new set of rules constitutes a new game.[20]

Another concern one might have is that it may be impossible to explicitly formulate the rules. Thus, Brouwer thought that the rules of intuitionist reasoning could not be formally circumscribed. Or one might hold that Gödel's first incompleteness theorem shows that the rules for arithmetic cannot be explicitly formulated, since the theory is not axiomatic. The particular examples raise many interesting issues. Maybe Brouwer was shown to be wrong when Heyting axiomatised intuitionist logic. Maybe the axiomatisability of arithmetic is to be accommodated by the possibility that arithmetic is inconsistent (and complete). But whatever the case with the particular examples, there is no real worry here. There is no reason why, in general, the rules of some practice (a game, a branch of mathematics) *must* be explicitly formulable. The question is only: can we follow them? If there are cases where the rules cannot be explicitly formulated (perhaps because they use second-order logic), this means, presumably, that the rules cannot be taught by giving them to another person explicitly. But maybe in such cases, the rules are, in some sense, hard-wired within us—in the way in which, according to Chomsky, a universal grammar is hard-wired in us—and in virtue of which, after sufficient prompting, we just "catch on".

A more worrying objection is posed by the spectre of relativism. What has happened to truth on this picture? The answer is as follows. Distinguish between pure and applied mathematics. What we have been

---

[20]Lurking in the background here is the thorny question of how to individuate practices. What are the synchronic and diachronic identity conditions for these? Arguably, a minor modification of the rules of test cricket leaves it as cricket, but one-day cricket and test cricket are not the same game, even though the one evolved from the other. Fortunately, we do not need to try to resolve these matters here.

talking about so far is pure mathematics: proving theorems according to some set of rules. And here there is a relativity of truth. What is acceptable or unacceptable is defined by the rules of the practice itself. The criteria of truth are *internal* to the practice. This is relativism of sorts, but not, as far as I can see, a worrying one. In particular, it is no threat to objectivity. From within a practice, results will be objectively right or wrong, just as, within a game, a move is objectively legitimate or not. You just have to be clear what the practice in question is.

But in that case, what are we to make of people who argue that some bits of mathematics are just plain wrong? For example, Brouwer and similar intuitionists famously held that classical mathematicians had got it wrong; and paraconsistent set-theorists have argued that ZF(C) gets our set theory wrong. Since classical mathematics, intuitionist mathematics, paraconsistent set theory, and ZF(C) are all equally legitimate games, are such disputes simply mistaken?

No. What is at issue here is this. In each case there is a received practice, number theoretic reasoning, set theoretic reasoning, or whatever. But there can be legitimate disputes about what, exactly, the correct norms of that practice are. We formulate different sets of rules, trying to capture these. There can be a fact of the matter about who, if anyone, gets it right. In the same way, linguists can take a spoken language and try to formulate a set of rules which capture its grammar. Some grammars can just be wrong. Of course, once a set of formal rules is set up, they do characterise some language or other; and even if it is not the one targeted, it can still be spoken. Similarly, once rules for a mathematical practice are explicitly formulated, it can be followed. Thus, an advocate of paraconsistent set theory with unlimited comprehension does not have to claim that ZF(C) is wrong. ZF(C) is just as good a practice (*qua* practice) as paraconsistent set theory. It is just that those who adhere to it are wrong if they claim that it correctly characterises our naive practice about sets.

6 Applied Mathematics

Let us turn from pure mathematics to applied mathematics. The matter of relativism is quite different there. We apply mathematics for many purposes: to simplify electronic circuits, to compute the orbits of satellites, to test biological hypotheses statistically.[21] It is always an

---

[21] One can think of metamathematics as an applied mathematics too. Formulas are just finite strings of symbols, '∧', '∃', etc.; and axiom systems are collections of these, structured in a certain way. A metatheory establishes certain results about these.

important question as to which mathematical theory is to be used in each application. Sometimes, a branch of pure mathematics may arise out of an application: arithmetic was presumably like this. Sometimes, a branch of mathematics may be developed with an eye on an application: the infinitesimal calculus was like this. Sometimes, when we want to apply mathematics, we can take a pre-existing system (which may, before that, have had no application) off the shelf: group theory was like that for modern physics. But always there is an important conceptual distinction between the mathematical theory itself and its application to this, that, or the other.

And once we apply a mathematical theory, there are criteria of correctness external to the pure practice. We need to get the *right* mathematical theory for the application in question. What makes it right? Here the story depends on one's philosophy of science. If one is an instrumentalist, one cares for nothing save the directly testable consequences. The mathematical theory is right if, when one applies it, the predictions check out. That's all there is to the matter. If one is a realist, success of this kind will be a (fallible) mark of something deeper. The mathematics must describe what is really there. What this means is that, under the correlation deployed in the application (for example, between points in Euclidean space and space-time points, or between functions in a complex Hilbert space and quantum states *in re*), the mathematical structure and the physical structure are isomorphic. That is why facts about the domain of application can be read off from facts about the mathematical structure.[22]

Speaking of applications, one should note what seems to be a significant disanalogy between pure mathematics and games: there is, as far as I can see, no phenomenon of the application of games similar to the application of mathematics.[23] A natural question is: why? Why can mathematics be applied in a way that games cannot? Part of the answer is that mathematical practice deals with things that are propositional (and so truth-apt—or better, truth-in-a-structure-apt); games (generally) do not. That we are dealing with propositional objects allows a certain kind of application. This cannot be the whole answer, however. Writing fiction is also dealing with propositional objects. Yet this has no application similar to mathematics either. Why is this? Presumably, a large part of the answer is that the origin of much pure mathematics (arithmetic, geometry, calculus) was in some application or other, as I have already noted. Unsurprisingly, then, mathematics has applications:

---

[22]See Priest (2005, §7.8).

[23]Games can be applied, of course, to develop fitness, problem-solving ability, team spirit, or wot not. But this would seem to be quite a different matter.

it was designed to do so. But what of those parts of mathematics which were not designed with an eye to application? There would seem to be no *a priori* reason why such parts of mathematics sometimes find application. Perhaps we just have to accept this as a contingent feature of the world in which we live.[24] Perhaps, if we develop enough pure mathematical systems, some of them are bound to find application sooner or later.

7 Matters Ontological

Let me end by commenting on the ontological story about mathematics that goes with the picture I have painted. In fact, many ontological stories fit happily with it. For example, one might be a plenitudinous, or "really full-blooded" platonist.[25] Every theory characterises a domain of existent objects—not just the one and only one on which mathematical platonists are usually fixated. Alternatively, and just as plausibly, one might endorse a conventionalism, of the kind advocated by Carnap in 'Empiricism, Semantics, and Ontology'.[26] According to this view, questions of existence have meaning within, and only within, a linguistic framework (practice), and are settled by the rules thereof. Outwith such a context, questions of existence are meaningless.

Another possibility which goes very well with the view, and the one which I, in fact, prefer, is the noneist position sketched in *Towards Non-Being*.[27] A mathematics practice can be taken to characterise an object, or collection of objects. This characterisation is guaranteed to be true, not necessarily at the actual world, but at some world or other. The world may be impossible, however, in the sense that its logic may be different from that of the actual (and other possible) worlds. Thus, the pursuance of the practice can be seen as the exploration of the structure of some non-existent objects—or, at least, objects that do not exist at this world. All games are, then, equally legitimate, in the sense that they all capture the way things are at some world (or worlds).

According to this view, mathematics and fiction are very similar activities. Mathematical theories (practices) and stories are free creations of the human spirit, and we can invent whatever we like. Having done so, we may then follow the inferential rules in play, to discover more about the mathematical or fictional situation characterised.[28] One can,

---

[24] See, e.g., Wigner (1960).
[25] As mooted in Beall (1999)
[26] Carnap (1950).
[27] Priest (2005, ch. 7).
[28] See, further, Priest (2005, §7.7).

if one likes, think of mathematical assertions or fictional assertions as coming prefixed with a tacit 'In the practice/story, it is the case that ...', just as we can think of legal assertions as prefixed by 'In such and such jurisdiction ...'. But of course, we are so used to operating certain practices, or of operating within a certain jurisdiction, that the prefix may become invisible to us.

This ontological position will answer another question that is likely to come up in connection with the story I have told. The view is mathematical pluralism. Does it entail a logical pluralism? Yes and no. Given the perspective of *Towards Non-Being*, worlds are many; and logic may differ from world to world. In that sense, yes. But there is only one actual world, only one actual truth, and so only one true logic. In that sense, no.

8 Conclusion

Like it or not, mathematical pluralism seems to be a fact of mathematical life—both diachronic and synchronic. Ours it is to make sense of this fact. One can, if one wishes, declare that there is one true mathematics (ZF(C)?), and that the rest is all mistaken. Such would seem to be a procrustean position of desperate proportion—and one, moreover, with a good deal of hubris. The position I have sketched is, I hope, much more plausible than this. There is a genuine plurality of mathematical practices—a motley, as Wittgenstein puts it[29]—each with its own set of rules. We are free to pursue any of them. All practices are equal. Though, of course, in terms of intrinsic interest, richness, beauty, application, etc., there will be significant differences. Some animals will always be more equal than others.

Acknowledgements

Versions of this paper have been given at various gatherings: the conference *This Conference Has No Name*, held at the Graduate Center, City University of NY, December 2009; the Melbourne Logic Group, March, 2010; an Arché FLC Seminar, University of St Andrews, May 2010; the University of Auckland, July 2010; and McGill University, October, 2010. I am grateful to the members of those audiences for their thoughtful comments, which improved the paper greatly. I am similarly grateful to a number of anonymous referees for this publication. The

---

[29]'I should like to say: mathematics is a MOTLEY of techniques of proof.—And upon this is based its manifold applicability and its importance.' (Wittgenstein 1978, III, 46).

final draft of this article was produced in the light of these, at a time when I had a heavy load of other commitments. I have no doubt that more time to reflect would have produced a smoother more polished paper.

References

Beall, J. C. (1999). "From Full Blooded Platonism to Really Full Blooded Platonism". In: *Philosophia Mathematica* 7, pp. 322–325.

Bridges, D. and F. Richman (1987). *Varieties of Constructive Mathematics*. Cambridge: Cambridge University Press.

Brown, B. and G. Priest (2004). "Chunk and Permeate, a Paraconsistent Inference Strategy; Part I, the Infinitesimal Calculus". In: *Journal of Philosophical Logic* 22, pp. 379–388.

Carnap, R. (1937). *The Logical Syntax of Language*. London: Kegan and Paul.

— (1950). "Empiricism, Semantics and Ontology". In: *Revue Internationale de Philosophie* 4, pp. 20–40.

Curry, H. (1958). *Outlines of a Formalist Philosophy of Mathematics*. Amsterdam: North-Holland.

Davies, E. B. (2005). "A Defence of Mathematical Pluralism". In: *Philosophia Mathematica* 13, pp. 252–276.

Dummett, M. (1977). *Elements of Intuitionism*. Oxford: Oxford University Press.

Hellman, G. and J. Bell (2006). "Pluralism and the Foundations of Mathematics". In: *Scientific Pluralism. Minnesota Studies in Philosophy of Science XIX*. Ed. by C. K. Waters, H. Longino, and S. Kellert. Minneapolis, MN: University of Minnesota Press.

Horsten, L. (2007). "Philosophy of Mathematics". In: *Stanford Encyclopedia of Philosophy*. Ed. by E. N. Zalta. URL: http://plato.stanford.edu/entries/philosophy-mathematics/(accessed_Nov_2011).

Koellner, P. (2011). *Truth in Mathematics: the Question of Pluralism*.

Mortensen, C. (1995). *Inconsistent Mathematics*. Dordrecht: Kluwer Academic Publishers.

— (2002). "Towards a Mathematics of Impossible Pictures". In: *Paraconsistency: the Logical Way to the Inconsistent*. Ed. by W. Carnielli, M. Coniglio, and I. D'Ottaviano. New York: Marcel Dekker, pp. 445–454.

Priest, G. (2005). *Towards Non-Being*. Oxford: Oxford University Press.

— (2006). *In Contradiction*. 2nd. Oxford: Oxford University Press.

— (2008). *Introduction to Non-Classical Logic: From If to Is*. Cambridge: Cambridge University Press.

Sambin, G. (2011). "A Minimalist Foundation at Work". In: *Logic, Mathematics, Philosophy, Vintage Enthusiasms: Essays in Honour of John L. Bell*. Ed. by D. DeVidi, M. Hallett, and P. Clarke. Heidelberg: Springer. Chap. 4.

Sweeney, D. J. (2013). "Chunk and Permeate: The Infinitesimals of Isaac Newton". In: *History and Philosophy of Logic*. URL: DOI:10.1080/01445340.2013.835099.

Vickers, P. (2007). "Was the early Calculus an Inconsistent Theory?" In: *PhilPapers*. URL: http://philpapers.org/rec/VICWTE.

Weber, Z. (2010). "Transfinite Numbers in Paraconsistent Set Theory". In: *Review of Symbolic Logic* 3, pp. 1–22.

Weir, A. (2011). "Formalism in the Philosophy of Mathematics". In: *Stanford Encyclopedia of Philosophy*. Ed. by E. N. Zalta. URL: http://plato.stanford.edu/entries/formalism-mathematics/ (accessed_Nov_2011).

Wigner, E. (1960). "The Unreasonable Effectiveness of Mathematics in the Natural Sciences". In: *Communications on Pure and Applied Mathematics* 13, pp. 1–14.

Wittgenstein, L. (1978). *Remarks on the Foundations of Mathematics*. revised. Oxford: Basil Blackwell.

CHAPTER 6

On Play-Objects in Dialogical Games. Towards A Dialogical Approach to Constructive Type Theory*

Shahid Rahman, Nicolas Clerbout and Zoe McConaughey

1 Introduction

In his 2011 book *The Interactive Stance*, J. Ginzburg stresses the utmost importance of taking conversational (interactive) aspects into account in order to develop a theory of meaning. Relying on various examples of actual conversations, Ginzburg argues that a significant part of meaning cannot be accounted for if one ignores features which have a typical interactive aspect. From this Ginzburg concludes that "grammar must link up directly with interaction" and describes one of the main purposes of his work, which is to give a formal framework suitable for (i) writing

---

*Jean Paul Van Bendegem published two papers on dialogical logic; one in the Synthese volume *New perspectives in dialogical logic* —Van Bendegem (2001)— as an answer to Rahman and Carnielli (2000), and another with Rahman on paraconsistency and adaptive logic—Rahman and Van Bendegem (2002).

We thought that the following paper for Jean Paul's Festschrift should provide a new impulse to continue the exploration of dialogical logic within the context of a subject that is in the scope of his own researches, namely: the bridges between philosophy of mathematics and dynamic approaches to a theory of meaning.

rules which account for the import of linguistic acts in a conversation and (ii) giving conversational rules that define the conditions under which different kinds of linguistic acts are possible in different kinds of conversational contexts.[1] For this purpose, Ginzburg chooses to use the formalism of Type Theory (with records). The formalism, in particular in its constructive version, has proven to be successful when applied to the non-conversational part in natural language.[2] These achievements of CTT ("constructive type theory") are certainly a strong motivation for Ginzburg's choice. Nevertheless in this paper we propose an alternative approach.

The dialogical approach to meaning is based on the idea that expressions get their meaning through the way they are used in certain kinds of dialogue games between a Proponent and an Opponent. It has been applied successfully to a variety of formal languages of interest in logic.[3] Thus, the dialogical framework is a promising alternative candidate for a general theory of meaning which meets Ginzburg's aims. However, very little has been done when it comes to applying this framework to natural languages. Indeed a lot remains to be done in order to develop a general dialogical theory of meaning. Our aim here is to make a first incursion in this field. For this purpose we introduce what we call play-objects and design dialogical games for an explicit language in which these objects occur.

The games we describe are clearly inspired by CTT. In particular, they rely on some principles on which Martin-Löf's formalism is grounded such as the "propositions as sets" principle, the "no entity without a type" principle, etc.[4] Enriching standard dialogues in accordance with these ideas is a necessary step in order to establish connections between the dialogical approach and CTT. However it would be a mistake to conclude that we obtain nothing more than a dialogical re-writing of CTT. Indeed, as we extensively discuss in Section 2, there is an important difference between dialogical play-objects and CTT's proof-objects, and this difference cannot easily be overcome. This is not surprising because we introduce play-objects so that interaction can be explicitly accounted for within dialogue games, while interactive aspects are clearly not covered by CTT's proof-objects. Summing up: introduc-

---

[1] Ginzburg (2011, pp. 5, 9).
[2] See in particular Ranta (1994).
[3] See for example Rahman and Keiff (2005), Keiff (2009), Redmond and Fontaine (2011). The dialogical semantics for standard first-order language is recalled at the end of this work.
[4] Cf. Martin-Löf (1984). For updated presentations and discussions see Granström (2011), Nordström, Petersson, and Smith (1990), Primiero (2008) and Ranta (1994).

ing play-objects is a necessary step for a dialogical approach to CTT and in order to provide a theory of meaning which not only is based on interactive aspects but also explicitly accounts for interaction. But it is not enough to give a detailed account of the connections between dialogues and CTT. These connections are likely to be found when we move from dialogues to strategies. In Section 3 we take an example of a dialogue from which we introduce various matters related to the level of strategies. We only sketch possible directions for future works, though, because the questions we raise require a careful and detailed analysis, a work still in progress.

In addition to the topic of the relation between dialogues and CTT, the elements we mention in Section 3 are also of interest for another question, one which is more peculiar to the study of the dialogical framework itself. When it comes to the study of the level of strategies, in works dedicated to the dialogical framework, almost all contributions are limited to considerations about the existence of certain strategies for the players, while a lot more can and should be considered. An important task, among others, is to give a precise way to describe strategies, how they can be composed, etc. The remarks we make in Section 3 look promising for the task of contributing to this task. To be more specific, we suggest a possible way to use play-objects for such purposes by using or adapting some features of CTT, in particular the computational rules which can be used in CTT to give "traces of proofs" – see Thompson (1991).

2 Play-objects and Particle rules

In the framework of constructive type theory propositions are sets whose elements are called proof-objects. When such a set is not empty, it can be concluded that the proposition has a proof and that it is true. In his 1988 paper, Ranta proposed a way to make use of this approach in relation to game-theoretical approaches. Ranta took Hintikka's Game Theoretical Semantics as a case study, but the point does not depend on this particular framework. Ranta's idea was that in the context of game-based approaches, a proposition is a set of winning strategies for the player positing the proposition.[5] Now in game-based approaches, the notion of truth is to be found at the level of such winning strategies. This idea of Ranta's should therefore enable us to apply safely and directly methods taken from constructive type theory to cases of game-based approaches.

But from the perspective of game theoretical approaches, reducing a

---

[5]That player can be called Player 1, Myself or Proponent.

game to a set of winning strategies is quite unsatisfactory, all the more when it comes to a theory of meaning. This is particularly clear in the dialogical approach in which different levels of meaning are carefully distinguished. There is thus the level of strategies which is a level of meaning analysis, but there is also a level prior to it which is usually called the level of plays. For this reason we would rather have propositions interpreted as sets of what we shall call *play-objects*, reading an expression

$$p : \varphi$$

as "p is a play-object for φ".

2.1 The Formation of Propositions

Before delving into the details about play-objects, let us first discuss the issue of the formation of expressions and in particular of propositions.

In standard dialogical systems, there is a presupposition that the players use well-formed formulas. One can check the well formation at will, but only with the usual meta reasoning by which one checks that the formula indeed observes the definition of wff. The first enrichment we want to make is to allow players to question the status of expressions, in particular to question the status of something as actually standing for a proposition. This is inspired by the formation rules of the CTT framework. Thus, we start with rules giving a dialogical explanation of the *formation* of propositions. These are local rules added to the particle rules of Section 2.2 which give the local meaning of logical constants. "Formation" dialogues have been introduced in Rahman (2012) and further developed in Rahman and Clerbout (2013). It is necessary to have them not only for the sake of studying the connections between our games and the CTT framework, but also for our general project of a theory of meaning based on interaction.

Let us make a remark before displaying the formation rules. Because the dialogical theory of meaning is based on argumentative interaction, dialogues feature expressions which are not posits of sentences, namely: requests used for challenges. This is illustrated by the formation rules below and the particle rules in the next section. By the *no entity without type* principle, the type of these actions, which we may call "*formation-request*", should be specified during a dialogue. We shall consider that the force symbol $?_F$ already makes the type explicit. The fact that the force symbol $?_F$ in itself makes the type *request* explicit should not be confused with a move which posits that an entity is of the type requested.[6] Requests are written according the rules in the table below.

Posit	Challenge[†]	Defence
$\mathbf{X}\,!\,\Gamma$: set	$\mathbf{Y}\,?_{can}\,\Gamma$ or $\mathbf{Y}\,?_{gen}\,\Gamma$ or $\mathbf{Y}\,?_{eq}\,\Gamma$	$\mathbf{X}\,!\,a_1:\Gamma,\mathbf{X}\,!\,a_2:\Gamma,...$ ($\mathbf{X}$ gives the canonical elements of Γ) $\mathbf{X}\,!\,a_i:\Gamma \to a_j:\Gamma$ ($\mathbf{X}$ provides a generation method) $\mathbf{X}$ gives the equality rule for Γ [‡]
$\mathbf{X}\,!\,\varphi \vee \psi$: prop	$\mathbf{Y}\,?_{F\vee 1}$ or $\mathbf{Y}\,?_{F\vee 2}$	$\mathbf{X}\,!\,\varphi$: prop respectively $\mathbf{X}\,!\,\psi$: prop
$\mathbf{X}\,!\,\varphi \wedge \psi$: prop	$\mathbf{Y}\,?_{F\wedge 1}$ or $\mathbf{Y}\,?_{F\wedge 2}$	$\mathbf{X}\,!\,\varphi$: prop respectively $\mathbf{X}\,!\,\psi$: prop
$\mathbf{X}\,!\,\varphi \to \psi$: prop	$\mathbf{Y}\,?_{F\to 1}$ or $\mathbf{Y}\,?_{F\to 2}$	$\mathbf{X}\,!\,\varphi$: prop respectively $\mathbf{X}\,!\,\psi$: prop
$\mathbf{X}\,!\,(\forall x:A)\varphi(x)$: prop	$\mathbf{Y}\,?_{F\forall 1}$ or $\mathbf{Y}\,?_{F\forall 2}$	$\mathbf{X}\,!\,A$: set respectively $\mathbf{X}\,!\,\varphi(x)$: prop $(x:A)$
$\mathbf{X}\,!\,(\exists x:A)\varphi(x)$: prop	$\mathbf{Y}\,?_{F\exists 1}$ or $\mathbf{Y}\,?_{F\exists 2}$	$\mathbf{X}\,!\,A$: set respectively $\mathbf{X}\,!\,\varphi(x)$: prop $(x:A)$
$\mathbf{X}\,!\,B(k)$: prop (for atomic B)	$\mathbf{Y}\,?_F$	$\mathbf{X}\,sic\,(n)$ ($\mathbf{X}$ indicates that $\mathbf{Y}$ posited it in move n)
$\mathbf{X}\,!\,\bot$: prop	$--$	$--$

†: When different challenges are possible, the challenger chooses.
‡: See a presentation of equality rules in Appendix 2.

By definition the falsum symbol $\bot$ is of type prop. A posit $\bot$ cannot therefore be challenged.

The next rule is not a formation rule *per se* but rather a substitution rule.[7] When φ is an elementary sentence, the substitution rule helps explaining the formation of such sentences.

POSIT-SUBSTITUTION

There are two cases in which $\mathbf{Y}$ can ask $\mathbf{X}$ to make a substitution in the context $x_i:A_i$. The first one is when in a standard play a (list of) variable occurs in a posit with a proviso. Then the challenger posits an instantiation of the proviso.

Posit	Challenge	Defence
$\mathbf{X}\,!\,\pi(x_1,\ldots,x_n)\,(x_i:A_i)$	$\mathbf{Y}\,!\,\tau_1:A_1\ldots\tau_n:A_n$	$\mathbf{X}\,!\,\pi(\tau_1,\ldots,\tau_n)$

The second case is in a formation-play. In such a play the challenger

---
[6]Such a move could be written as $?_{F\vee 1}$: *formation-request*.
[7]It is an application of the original rule from CTT given in Ranta (1994, p. 30).

simply posits the whole assumption as in move 7 of the example below:

Posit	Challenge	Defence
X ! $\pi(\tau_1, \ldots, \tau_n) \, (\tau_i : A_i)$	**Y** ! $\tau_1 : A_1 \ldots \tau_n : A_n$	**X** ! $\pi(\tau_1, \ldots, \tau_n)$

By way of illustration of the formation rules, we present a dialogue where the Proponent posits the thesis $(\forall x : A)B(x) \to C(x)$: prop given that A: set, $B(x)$: prop$(x : A)$ and $C(x)$: prop$(x : A)$, where the three provisos appear as initial concessions by the Opponent.[8] Good form demands that we first present the structural rules which define the conditions under which a play can start, proceed and end. But we leave these rules for the next section: they are not needed to understand the example.

	O				**P**	
I	! A : set					
II	! $B(x)$: prop $(x : A)$					
III	! $C(x)$: prop $(x : A)$					
					! $(\forall x : A)B(x) \to C(x)$: prop	0
1	n := 2				m := 2	2
3	$?_{F\forall 1}$	(0)			! A : set	4
5	$?_{F\forall 2}$	(0)			! $B(x) \to C(x)$: prop $(x : A)$	6
7	! $x : A$	(6)			! $B(x) \to C(x)$: prop	8
9	$?_{F\to 1}$	(8)			! $B(x)$: prop	12
11	! $B(x)$: prop		(II)		! $x : A$	10
13	$?_{F\to 2}$	(8)			! $C(x)$: prop	16
15	! $C(x)$: prop		(II)		! $x : A$	14

Explanations:

I to III: **O** concedes that A is a set and that $B(x)$ and $C(x)$ are propositions provided x is an element of A,

Move 0: **P** posits that the main sentence, universally quantified, is a proposition (under the concessions made by **O**),

Moves 1 and 2: the players choose their repetition ranks,

Move 3: **O** challenges the thesis a first time by asking the left-hand part as specified by the formation rule for universal quantification,

Move 4: **P** responds by positing that A is a set. This has already been granted with the premise I so **P** can make this move while respecting the Formal rule,

---

[8] The example comes from Ranta (1994, p. 31).

Move 5: **O** challenges the thesis again, this time asking for the right-hand part,[9]

Move 6: **P** responds, positing that $B(x) \to C(x)$ is a proposition provided $x : A$,

Move 7: **O** uses the substitution rule to challenge move 6 by granting the proviso,

Move 8: **P** responds by positing that $B(x) \to C(x)$ is a proposition,

Move 9: **O** then challenges move 8 a first time by asking the left-hand part as specified by the formation rule for material implication.

In order to defend **P** needs to make an elementary move. But since **O** has not played it yet, **P** cannot defend at this point. Thus:

Move 10: **P** launches a counterattack against assumption II by applying the first case of the substitution rule,

Move 11: **O** answers move 10 and posits that $B(x)$ is a proposition,

Move 12: **P** can now defend in reaction to move 9,

Move 13: **O** challenges move 8 a second time, this time requiring the right-hand part of the conditional,

Move 14: **P** launches a counterattack and challenges assumption III by applying again the first case of the substitution rule,

Move 15: **O** defends by positing that $C(x)$ is a proposition,

Move 16: **P** can now answer to the request of move 13 and win the dialogue (**O** has no further move).

From the view point of building a winning strategy, the Proponent's victory only shows that the thesis is justified *in this particular play*. To build a winning strategy we must run all the relevant plays for this thesis under these concessions.

Now that the dialogical account of formation rules has been clarified, we may develop further our analysis of plays by introducing play-objects.

---

[9]This can be done because **O** has chosen 2 for her repetition rank.

2.2 Particle rules

The idea is now to design dialogical games in which the players' posits are of the form "$p : \varphi$" and acquire their meaning in the way they are used in the game – i.e., how they are challenged and defended. This requires, among others, to analyse the form of a given play-object p, which depends on φ, and how a play-object can be obtained from other, simpler, play-objects.

The standard dialogical semantics[10] for logical constants gives us the needed information for this purpose. The main logical constant of the expression at stake provides the basic information as to what a play-object for that expression consists of:

A play for **X**! $\varphi \vee \psi$ is obtained from two plays p_1 and p_2, where p_1 is a play for **X**! φ and p_2 is a play for **X**! ψ. According to the particle rule for disjunction, it is the player **X** who can switch from p_1 to p_2 and vice-versa. To show this, we write that the play is of the form $(p_1 + p_2)$.

A play for **X**! $\varphi \wedge \psi$ is obtained similarly, except that it is the player **Y** who can switch from p_1 to p_2. To show this, we write that the play is of the form $(p_1 \otimes p_2)$.

A play for **X**! $\varphi \to \psi$ is obtained from two plays p_1 and p_2, where p_1 is a play for **Y**! φ and p_2 is a play for **X**! ψ. It is the player **X** who can switch from p_1 to p_2. We write that the play is of the form $(p_1 \multimap p_2)$.

The standard dialogical particle rule for negation rests on the interpretation of $\neg \varphi$ as an abbreviation for $\varphi \to \bot$, although it is usually left implicit. It follows that a play for **X**! $\neg \varphi$ is also of the form $(p_1 \multimap p_2)$, where p_1 is a play for **Y**! φ and p_2 is a play for **X**! $\bot$, where **X** can switch from p_1 to p_2.

Notice that this approach covers the standard game-theoretical interpretation of negation as role-switch: p_1 is a play for a **Y** move.

As for quantifiers, we give a detailed discussion after the particle rules (see next page).[11] For now, we would like to point out that, just like what is done in constructive type theory, we are dealing with quantifiers for which the type of the bound variable is always specified. We thus consider expressions of the form $(Qx : A)\varphi$, where Q is a quantifier symbol.

---

[10] See Appendix 1.

[11] A similar comment to the one we made for formation-requests on p. 3 can be made here.

It may seem unfortunate that we use symbols that are usually used to denote linear connectives ($\otimes$, $\multimap$). We use these because their game-theoretical interpretations (Blass 1992) completely match the descriptions we have just given of how play-objects can be obtained from simpler ones. Notice that we have added for each logical constant a challenge of

Posit	Challenge	Defence
X ! φ (where no play-object has been specified for φ)	**Y** ?*play-object*	**X** ! $p : \varphi$
	Y ?$_{prop}$	**X** ! $\varphi \vee \psi$: prop
X ! $p : \varphi \vee \psi$	**Y** ?$[\varphi, \psi]$	**X** ! $L^\vee(p) : \varphi$ or **X** ! $R^\vee(p) : \psi$
Description of $p : (p_1 + p_2)$		[the defender has the choice]
	Y ?$_{prop}$	**X** ! $\varphi \wedge \psi$: prop
	Y ?$[\varphi]$	**X** ! $L^\wedge(p) : \varphi$
X ! $p : \varphi \wedge \psi$	or	respectively
	Y ?$[\psi]$	**X** ! $R^\wedge(p) : \psi$
Description of $p : (p_1 \otimes p_2)$	[the challenger has the choice]	
	Y ?$_{prop}$	**X** ! $\varphi \rightarrow \psi$: prop
X ! $p : \varphi \rightarrow \psi$	**Y** ! $L^\rightarrow(p) : \varphi$	**X** ! $R^\rightarrow(p) : \psi$
Description of $p : (p_1 \multimap p_2)$		
	Y ?$_{prop}$	**X** ! $\neg\varphi$: prop
X ! $p : \neg\varphi$	**Y** ! $L^\rightarrow(p) : \varphi$	**X** ! $R^\rightarrow(p) : \bot$
Description of $p : (p_1 \multimap \bot)$		
	Y ?$_{prop}$	**X** ! $(\exists x : A)\varphi$: prop
	Y ?$_L$	**X** ! $L^\exists(p) : A$
X ! $p : (\exists x : A)\varphi$	or	respectively
	Y ?$_R$	**X** ! $R^\exists(p) : \varphi(L^\exists(p))$
Description of $p : (p_1 \otimes p_2)$	[the challenger has the choice]	
	Y ?$_L$	**X** ! $L^{\{\}}(p) : A$
X ! $p : \{x : A \mid \varphi\}$	or	respectively
	Y ?$_R$	**X** ! $R^{\{\}}(p) : \varphi(L^{\{\}}(p))$
Description of $p : (p_1 \otimes p_2)$	[the challenger has the choice]	
	Y ?$_{prop}$	**X** ! $(\forall x : A)\varphi$: prop
X ! $p : (\forall x : A)\varphi$	**Y** ! $L^\forall(p) : A$	**X** ! $R^\forall(p) : \varphi(L^\forall(p))$
Description of $p : (p_1 \multimap p_2)$		
	Y ?$_{prop}$	**X** ! $B(k)$: prop
X ! $p : B(k)$ (for atomic B)	**Y** ?	**X** $sic(n)$ (**X** indicates that **Y** posited it in move n)

the form "**Y**?$_{prop}$" by which the challenger questions the fact that the expression at the right-hand side of the semi-colon is a proposition. This makes the connection with the formation rules given in Section 2.1 *via* **X**'s defence. More details are given in the discussion after the structural rules.

It may happen, as we shall see in our example in Section 3, that the form of play-objects is not explicit at first. In such cases we deal with expressions of the form, e.g., "$p : \varphi \wedge \psi$". We may then use expressions of the form $L^\wedge(p)$ and $R^\wedge(p)$ – which we call *instructions* – in the relevant defences. Their respective interpretations are "take the left part of p" and "take the right part of p". In instructions we indicate the logical constant at stake. First it keeps the formulations explicit enough, in particular in the case of embedded instructions. More importantly we must keep in mind that there are important differences between play-objects depending on the logical constant. Remember for example that in the case of conjunction the play-object is a pair, but it is not in the case of disjunction. Thus $L^{\wedge}(p)$ and $L^{\vee}(p)$, say, are actually different things and the notation takes that into account.

Let us focus on the rules for quantifiers. Dialogical semantics highlights the fact that there are two distinct moments when considering the meaning of quantifiers: the choice of a value given to the bound variable, and the instantiation of the formula after replacing the bound variable with the chosen value. But at the same time in the standard dialogical approach there is some sort of presupposition that every quantifier symbol ranges on a unique kind of objects. Now, things are different in the context of the explicit language we borrow from CTT. Quantification is always relative to a set, and there are sets of many different kinds of objects (for example: sets of individuals, sets of pairs, sets of functions, etc). Thanks to the instructions we can give a general form for the particle rules. It is in a third, later, moment that the kind of object is specified, when instructions are "resolved" by means of the structural rule SR4.1 below.

Constructive type theory makes it clear that as soon as propositions are thought of as sets, there is a basic similarity between conjunction and existential quantifier on the one hand and material implication and universal quantifier on the other hand. Briefly, the point is that they are formed in similar ways and their elements are generated by the same kind of operations.[12] In our approach, this similarity manifests itself in the fact that a play-object for an existentially quantified expression is of the same form as a play-object for a conjunction. Similarly, a play-object for a universally quantified expression is of the same form as one for a material implication.[13]

The particle rule just before the one for universal quantification is a novelty in the dialogical approach. It involves expressions commonly

---

[12] More precisely, conjunction and existential quantifier are two particular cases of the Σ operator (disjoint union of sets), whereas material implication and universal quantifier are two particular cases of the Π operator (indexed product on sets). See for example Ranta (1994, Chapt. 2).

[13] Still, if we are playing with classical structural rules, there is a slight difference between material implication and universal quantification which we take from Ranta (1994, Table 2.3), namely that in the second case p_2 always depends on p_1.

used in constructive type theory to deal with separated subsets. The idea is to understand *those elements of A **such that** φ* as expressing that at least one element L(p) of A witnesses φ(L(p)). The same correspondence that linked conjunction and existential quantification now appears.[14] This is not surprising since such posits actually have an existential aspect: in $\{x : A \mid \varphi\}$ the left part "$x : A$" signals the existence of a play-object.

2.3 Play-objects versus proof-objects

In constructive type theory, four different kinds of rules are associated to logical operations: formation rules, introduction rules, elimination rules and computational rules.[15] The comparison between these rules and our particle rules stresses the fact that play-objects and type-theoretical proof-objects are fundamentally different, even though the former are strongly inspired by the latter.

We focus now on the task of comparing the rules given above with the other three kinds of rules: introduction rules, elimination rules and computational rules.

Particle rules are often considered to be the dialogical counterpart of the elimination rules of Natural Deduction. This comparison comes from the analytic aspect of the rules and of dialogues: defences are subformulas of the challenged formula. From this it is often believed that there is no dialogical counterpart for introduction rules. But these remarks are actually not accurate. A closer look at the connection between the dialogical approach and Natural Deduction shows that there is a connection between particle rules and both introduction and elimination rules. The connection is to be found when we look at particle rules as they are applied in the context of a strategy for the Proponent. This connection is the ground for giving a correspondence result between Natural Deduction proofs and the dialogical manifestation of validity. For a detailed analysis, see Rahman, Clerbout, and Keiff (2009).

To make the connection clear between particle rules on the one hand

---

[14]As pointed out in Martin-Löf (1984), subset separation is another case of the Σ operator. See in particular p. 53:
"Let A be a set and $B(x)$ a proposition for $x \in A$. We want to define the set of all $a \in A$ such that $B(a)$ holds (which is usually written $\{x \in A : B(x)\}$). To have an element $a \in A$ such that $B(a)$ holds means to have an element $a \in A$ together with a proof of $B(a)$, namely an element $b \in B(a)$. So the elements of the set of all elements of A satisfying $B(x)$ are pairs $(a; b)$ with $b \in B(a)$, i.e. elements of $(\Sigma x \in A)B(x)$. Then the Σ-rules play the role of the comprehension axiom (or the separation principle in ZF)."

[15]See Thompson (1991, Chapt. 4).

and introduction / elimination rules on the other hand, we have to leave the level of particular dialogues and consider the level of strategies (sets of dialogues). This is a manifestation of the difference between the meaning explanations given in CTT, which rest on the notion of proof-object, and the meaning explanations given by our particle rules. We have introduced play-objects because in dialogues, propositions are considered as sets of plays. This is clearly inspired from CTT, but should not lead to identify play-objects with proof-objects.

Such a confusion may probably result from the seeming similarity between play-objects and proof-objects when looking at winning conditions for conjunction and disjunction. However, at the level of the local meaning (provided by the particle rules) no winning conditions have yet been specified. One could think of particle rules as describing different kinds of request-response sequences – in a similar way as in the sequences generated by the notion of *arena* in game theory. The difference becomes obvious when we consider material implication (and universal quantification). We have noticed that a play-object for a formula $\varphi \rightarrow \psi$ is of the form $(p_1 \multimap p_2)$. But in the explanation for the sign "$\multimap$" and the local meaning provided by the particle rule there is no requirement of any particular link between p_1 and p_2: the play-object for the tail does not have to result from the play-object for the head by means of some kind of function. This plainly differs from the meaning explanation of such formulas in CTT, where a proof-object is of the form "$(\lambda x.p_2)p_1$": a function from the proof-objects for the head to the proof-object for the tail. Certainly, if we add intuitionistic structural rules, building a winning strategy will enforce a functional interpretation for the *proof* of a material implication, but this does not have bearings on the *local meaning* of this logical constant, displayed by the particle rules.

The difference between play-objects and proof-objects is consistent with the various remarks we made about the fact that the connection between CTT and game-based approaches is not as straightforward as Ranta assumes in his 1988 paper. Actually the difference is not surprising when considering the dialogical distinction between the levels of plays and of strategies. The level governed by particle rules can be thought of as preliminary. From the perspective of constructive type theory it would probably be considered as incomplete because it does not yet rest on the notion of proof. In our view, the relation between dialogical play-objects and CTT proof-objects is best understood in the following way: *play-objects are posited proof-objects*. We could then say that a given play-object actually stands for a proof-object if the posit is justified. In other words: if there is a winning strategy for the Proponent in the corresponding dialogical game. Such a view needs to be carefully explained and checked. In the next Section we give an example of a dialogue and to suggest various directions which need to be developed in order to give a full dialogical approach of CTT.

3 Example and further directions: from play-objects to strategies

In this Section we illustrate our enriched dialogical framework by giving a dialogue involving the famous donkey sentence. We also take the opportunity to make preliminary remarks on matters related to the level of strategies, which we will need to consider in the future to give a precise account of the relation between the dialogical and the type theoretical approaches.

Before that let us say a few words about the other kind of dialogical rules called structural rules. These rules govern the way plays globally proceed and are therefore an important aspect of a dialogical semantics. We work with the following structural rules:

SR0 (Starting rule) Any dialogue starts with the Proponent positing the thesis. After that the players each choose a positive integer called repetition ranks.

SR1i (Intuitionistic Development rule) Players move alternately. After the repetition ranks have been chosen, each move is a challenge or a defence in reaction to a previous move, in accordance with the particle rules. The repetition rank of a player bounds the number of challenges he can play in reaction to a same move. Players can answer only against *the last non-answered* challenge by the adversary.

SR1c (Classical Development rule) Players move alternately. After the repetition ranks have been chosen, each move is a challenge or a defence in reaction to a previous move, in accordance with the particle rules. The repetition rank of a player bounds the number of challenges and defences he can play in reaction to a same move.

SR2 (Formation first) O starts by challenging the thesis with the request "$?_{prop}$". The game then proceeds by applying the formation rules first in order to check that the thesis is indeed a proposition. After that the Opponent is free to use the other local rules insofar as the other structural rules allow it.

SR3 (Modified Formal rule) O's elementary sentences can not be challenged, however O can challenge an elementary sentence (posited by P) iff herself (the opponent) did not posit it before.

SR4.1 (Resolution of instructions) Whenever a player posits a move where instructions $I_1, \ldots, I_n$ occur, the other player can ask him to replace these instructions (or some of them) by suitable play-objects.

If the instruction (or list of instructions) occurs at the right of the colon and the posit is the tail of an universally quantified sentence or of an implication (so that these instructions occur at the left of the colon in the posit of the head of the implication), then it is the challenger who can choose the play-object – in these cases the player who challenges the instruction is also the challenger of the universal quantifier and/or of the implication.

Otherwise it is the defender of the instructions who chooses the suitable play-object. That is:

Posit	Challenge	Defence
$\mathbf{X} \,!\, \pi(I_1, \ldots, I_n)$	$\mathbf{Y}\; I_1, \ldots, I_m =?\; (m \leq n)$	$\mathbf{X} \,!\, \pi(b_1, \ldots, b_m)$

- If the instruction that occurs at the right of the colon is the tail of either a universal or an implication (such that $I_1, \ldots, I_n$ also occur at the left of the colon in the posit of the head), then $\mathbf{b_1}, \ldots, \mathbf{b_m}$ **are chosen by the challenger**.
- Otherwise **the defender chooses**.

Important remark. In the case of embedded instructions of the form $I_1(\ldots(I_k)\ldots)$, the substitutions are thought as being carried out from I_k to I_1: first substitute I_k with some play-object b_k, then $I_{k-1}(b_k)$ with $b_{k-1} \ldots$ until $I_1(b_2)$. If such a progressive substitution has actually been carried out once, a player can then replace $I_1(\ldots(I_k)\ldots)$ directly.

SR4.2 (Substitution of instructions) Whenever during play, play-object b has been chosen by any of both players for an instruction I, and player $\mathbf{X}$ posits $!\pi(I)$, then the antagonist can ask to substitute I with b in any posit $\mathbf{X}!\pi(I)$:

Posit	Challenge	Defence
Player 1 $!\; \pi_i(I)$		
Player 2 $!\; I =?$		
Player 3 $!\; \pi_i(b)$		
$\ldots$		
$\mathbf{X} \,!\, \pi_j(I)$	$\mathbf{Y}\; ?\; b/I$	$\mathbf{X} \,!\, \pi_j(b)$

The idea is that the resolution of an instruction in a move yields a certain play-object for some substitution term, and therefore the same play-object can be assumed to result for any other occurrence of the same substitution term: instructions are functions after all and as such they must yield the same play-object for the same substitution term.

In order to *quantify* into instructions $I^\vee$ – that is, either $L(x)$ or $R(x)$ – the following substitution rule is added:

Posit	Challenge	Defence
X ! $\pi(I^\vee(x))\,(x:A)$	**Y** ? $a:A/I^\vee$	**X** ! $\pi(a)$

Similar applies to $I^\wedge$ and $I^\exists$:

Posit	Challenge	Defence
X ! $\pi(L^\wedge(x \otimes p_2))\,(x:A)$	**Y** ? $a:A/L^\wedge$	**X** ! $\pi(a)$
X ! $\pi(R^\wedge(p_1 \otimes y))\,(y:B)$	**Y** ? $b:B/R^\wedge$	**X** ! $\pi(b)$
X ! $\pi(L^\exists(x \otimes p_2))\,(x:A)$	**Y** ? $a:A/L^\exists$	**X** ! $\pi(a)$
X ! $\pi(R^\exists(p_1 \otimes y))\,(y:B)$	**Y** ? $b:B/R^\exists$	**X** ! $\pi(b)$

SR5 (Winning rule for dialogues) For any p, a player who posits "$p :\bot$" looses the current dialogue. Otherwise the player who makes the last move in a dialogue wins it.

A detailed explanation of the standard rules can be found in Appendix 1. In the rules we just gave there are some additions, namely those numbered SR2 and SR4.1-2 here, and also the first part of the winning rule. Since we made explicit the use of $\bot$ in our games, we need to add a rule for it: the point is that positing *falsum* leads to immediate loss; we could say that it amounts to a withdrawal (Keiff 2007).

We need the rules SR4.1 and SR4.2 because of some features of CTT's explicit language. In CTT it is possible to account for questions of dependency, scope, etc. directly at the level of the language. In this way various puzzles, such as anaphora, get a convincing and successful treatment. The typical example, which we consider below, is the so-called *donkey sentence* "Every man who owns a donkey beats it". These two rules give a means for the players to associate and substitute play-objects for what we have called instructions. See the dialogue below for an application.

Notice that there is a principle from CTT that we did not entirely apply in this first paper, namely that "no entity comes without a type". Indeed SR0 introduces repetition ranks (to ensure finiteness of plays) and we have not said anything about their type. This is still a job to be done.

We now give an example of a dialogue for the donkey sentence. In his 1986 paper, G. Sundholm thoroughly discussed this famous puzzle. As he pointed out, the problem is to give a way to capture the back-reference of the pronoun "it" in the sentence "Every man who owns a donkey beats it". For that we first notice that "a man who owns a donkey" is an element of the set

$$\{x : M \mid (\exists y : D) Oxy\},$$

making use of subset separation. From there it is easy to use projections to get the following formula for the donkey sentence:

$$(\forall z : \{x : M \mid (\exists y : D) Oxy\}) B(L(z), L(R(z)))$$

where M is the set of men, D is the set of donkeys, Oxy stands for "x owns y" and Bxy stands for "x beats y". In this way we account for the fact that the pronoun "it" refers to the donkey mentioned in the first part of the sentence.

In the following dialogue, the donkey sentence is conceded together wih other posits by the Opponent. Given these concessions, the Proponent posits "$Beats(m,d)$" as the thesis.

		O				P	
I		$!p : (\forall z : \{x : M \mid (\exists y : D) Oxy\}) Beats(L(z), L(R(z)))$					
II		$!m : M$					
III		$!d : D$					
IV		$!p' : Omd$					
						$!Beats(m,d)$	0
1		$\mathbf{n} :=$				$\mathbf{m} :=$	2
3		?play-object	(0)			$!q : Beats(m,d)$	30
25		$!R^\forall(p) : Beats(L(z), L(R(z)))$		(I)		$!L^\forall(p) : \{x : M \mid (\exists y : D) Oxy\}$	4
5		$L^\forall(p) =?$	(4)			$!z : \{x : M \mid (\exists y : D) Oxy\}$	6
7		$?_L$	(6)			$!L^{\{\}}(z) : M$	8
9		$L^{\{\}}(z) =?$	(8)			$!m : M$	10
11		$?_R$	(6)			$!R^{\{\}}(z) : (\exists y : D) Omy$	12
13		$R^{\{\}}(z) =?$	(12)			$!(L^\exists(R^{\{\}}(z)), R^\exists(R^{\{\}}(z))) : (\exists y : D) Omy$	14
15		$L^\exists(R^{\{\}}(z)) =?, R^\exists(R^{\{\}}(z)) =?$	(14)			$!(d, p') : (\exists y : D) Omy$	16
17		$?_L$	(16)			$!L^\exists(d, p') : D$	18
19		$L^\exists(d, p') =?$	(18)			$!d : D$	20
21		$?_R$	(16)			$!R^\exists(d, p') : Omd$	22
23		$R^\exists(d, p') =?$	(22)			$!p' : Omd$	24
27		$!R^\forall(p) : Beats(m,d)$		(25)		$m/L^{\{\}}(z), d/L^\exists(R^{\{\}}(z))$	26
29		$!q : Beats(m,d)$		(27)		$Rp^\forall(p) =?$	28

Explanations. We left the repetition ranks unspecified (moves 1 and 2) and simply assume that they are big enough for **O** to play all her challenges and for **P** to answer. We also ignored redundant repetitions and focused on the steps which are relevant for the outcome of the play.

Now, because of the modified formal rule SR3 the Proponent must delay his answer to move 3. He thus counter-attacks by challenging **O**'s first concession (the donkey sentence). Then the Opponent has various choices:[16] in this dialogue she starts with a counterattack, asking **P** to choose a play-object for $L^\forall(p)$. The dialogue goes on with **O** playing in accordance with the particle rules and asking for resolutions of instructions during the process. Notice that when answering to challenge 13, the Proponent gives a description of $R^{\{\cdots\}}(z)$: it is a pair consisting of a left part and a right part. This allows him to introduce the instruction $L^\exists(R^{\{\cdots\}})(z)$ for the continuation of the play. With move 16 **P** chooses the play-objects d and p' as parts of the pair in order to use concessions III and IV at moves 20 and 24.

After move 24 the Opponent has no other choice but to answer move 4. Then it is easy for **P** to use rules SR4.2 and SR4.1 (with moves 26 and 28) in order to get exactly what he needs to play move 30 and win this dialogue.

Notice that as the dialogue unfurls, a more precise formulation of the initial play-object for the donkey sentence is revealed. In particular we obtained with moves 8, 12 and 14 important specifications on the form of $L^\forall(p)$. Using the notation we have introduced in Section 2, we get the following description for the play-object p:

$$L^{\{\cdots\}}(z) \otimes (L^\exists(R^{\{\cdots\}}(z)) \otimes R^\exists(R^{\{\cdots\}}(z)))) \multimap \mathit{Beats}(L^{\{\cdots\}}(z), L^\exists(R^{\{\cdots\}}(z))) \quad (A)$$

We can even keep track of which moves are played by **O** and by **P**. For this purpose we place the players' identities in the following way:

$$(L^{\{\cdots\}}(z) \otimes^{\mathbf{P}} (L^\exists(R^{\{\cdots\}}(z)) \otimes^{\mathbf{P}} R^\exists(R^{\{\cdots\}}(z)))) \multimap^{\mathbf{O}} \mathit{Beats}(L^{\{\cdots\}}(z), L^\exists(R^{\{\cdots\}}(z))) \quad (B)$$

We could borrow some terminology from constructive type theory and call expression (B) a trace or blue-print of the play for **O**'s concession I. Our interest in such expressions lies in the fact that we can see dialogues such as the one above as resulting from **P** following a certain strategy S. More precisely, the dialogue is one of those which can result when **P** plays according to S. An expression such as (B) can thus be considered as giving a partial description of the strategy S. Here, it is the part related to **O**'s first concession in this dialogue. It is not really clear yet how such a description of a strategy can be obtained from blue-prints

---

[16] The reader may check that **P** has a way to win no matter how the Opponent chooses to react to move 4. The hardest one is probably when **O** chooses to answer directly the challenge. In this case the trick for **P** is to choose the correct order in his moves and to use carefully the substitution rules given in rule SR4.2.

of plays, and a detailed analysis on this matter is mandatory. Actually everything remains to be done on this topic. In this work we can only point out what we believe is a promising starting point in order to give precise descriptions of strategies in terms of lists of instructions.

An arguable drawback of the approach we suggest is that descriptions such as (B) are heavy and difficult to handle. In spite of its length and the notation, the dialogue above is rather simple, so it is likely that for more complicated dialogues we will get very abstruse descriptions. What is more, the situation is likely to worsen when we combine such traces in order to give descriptions of strategies. An obvious way to make things better in this respect is to replace instructions by their associated play-objects. We could even consider this as mandatory since resolution of instructions is part of the dialogues. Hence, we could think of replacing (B) with the following:

$$(m \otimes^{\mathbf{O}} (m,d)) \multimap^{\mathbf{P}} Beats(m,d) \qquad (B')$$

Before discussing further this possibility, let us notice the following about the step from (B) to (B'). Notice that this step is reminiscent of the computational rules of constructive type theory. In CTT, elimination rules can be thought of as giving information on how to obtain a proof-object for a subformula given the proof-object of the starting formula. Computational rules then provide the means to compute the information in order to get the actual value of the proof-object for the subformula. For example one of the elimination rules for conjunction is

$$\frac{(p_1,p_2) : \varphi \wedge \psi}{fst(p_1,p_2) : \varphi} \quad (E1\wedge)$$

The fact that $fst(p_1,p_2)$ computes to p_1 is accounted for by the computational rule[17]

$$fst(p_1,p_2) \Rightarrow p_1$$

A possible way to deal with these rules from the dialogical perspective, which we leave open for further explorations, is the following. We can see such rules as special cases of the rule for functions. Just like propositions, we can give a particle rule for functions:

$$\mathbf{X} \,!\, f(x) : B\,(x : A) \;\big|\; \mathbf{Y} \,!\, a : A \;\big|\; \mathbf{X} \,!\, f(a) : B$$

The idea is that if we see computational rules as special cases of such functions, then we can implement them directly within dialogues as applied by the players. In this way we would obtain games where the

---

[17] See Thompson (1991, Chapt. 4).

players themselves can describe plays or strategies, because such descriptions are done by means of computational rules. This opens a new direction where we would have dialogues about plays and strategies, i.e., where we could develop a dialogical approach to the strategical level.

Besides simplifying the notation, there are various convincing reasons which make the use of CTT's computational rules desirable, or at least of something similar. Applying such a device could be a way to stress what different plays (or strategies) have in common. Again, the comparison with CTT is helpful to explain our point. The point is that in constructive type theory different proof processes (i.e., different derivations) can lead to identical proof-objects once the computational rules are applied.[18] From the dialogical point of view, a device of this kind is particularly relevant when we consider strategies. To take a very simple example, this could be the way to accurately formulate the basic similarity between different orders of moves and consider as basically similar the following two sequences of actions: "asking for the left conjunct then asking for the right conjunct" and "asking for the right conjunct then asking for the left conjunct".

Provided we define dialogical counterparts for the computational rules, the question remains whether we should systematically apply them and forget about non-simplified traces such as (B). In our view we should not consider that their only purpose is to apply simplification in order to get synthesized descriptions of strategies. Even though being able to account for similarities between plays or strategies is interesting, it is worth noticing that this would be achieved at the expense of a notation which keeps track of the players' actions. The point is the following: once we have replaced instructions by their values, like we do from (B) to (B'), we lose the chance to explicitly formulate strategies as lists of instructions. That is to say, there is something which is lost when we replace the instructions with their values, namely the way the value is settled. Such differences are precisely the ones which hide what different dialogues or strategies may have in common. When we are interested in such common aspects it is obviously better to apply simplifications and consider values instead of instructions.

Acknowledgements

The present paper is part of an ongoing project in the context of the MESHS Nord-Pas-de-Calais program "Argumentation, Decision, Action" (ADA).

---

[18]See Thompson (1991) Section 4.5.4 for an example and Section 11 for a discussion.

Appendix 1: Brief introduction to first-order dialogical games

Let L be a first-order language built as usual upon the propositional connectives, the quantifiers, a denumerable set of individual variables, a denumerable set of individual constants and a denumerable set of predicate symbols (each with a fixed arity).

We extend the language L with two labels **O** and **P**, standing for the players of the game, and the question mark "?". When the identity of the player does not matter, we use variables **X** or **Y** (with **X** ≠ **Y**). A *move* is an expression of the form "**X** − e", where e is either a formula φ of L or the form "?$[\varphi_1, ..., \varphi_n]$". We now present the rules of dialogical games. There are two distinct kinds of rules named particle (or local) rules and structural rules. We start with the particle rules.

Previous move	**X** − $\varphi \wedge \psi$	**X** − $\varphi \vee \psi$	**X** − $\varphi \to \psi$	**X** − $\neg \varphi$
Challenge	**Y**−?$[\varphi]$ or **Y**−?$[\psi]$	**Y**−?$[\varphi, \psi]$	**Y** − φ	**Y** − φ
Defence	**X** − φ resp. **X** − ψ	**X** − φ or **X** − ψ	**X** − ψ	−−

Previous move	**X** − $\forall x \varphi$	**X** − $\exists x \varphi$
Challenge	**Y**−?$[\varphi(a/x)]$	**Y**−?$[\varphi(a_1/x), \ldots, \varphi(a_n/x)]$
Defence	**X** − $\varphi(a/x)$	**X** − $\varphi(a_i/x)$ with $1 \leq i \leq n$

In this table, the a_i's are individual constants and $\varphi(a_i/x)$ denotes the formula obtained by replacing every occurrence of x in φ by a_i. When a move consists in a question of the form "?$[\varphi_1, ..., \varphi_n]$", the other player chooses one formula among $\varphi_1, ..., \varphi_n$ and plays it. We can thus distinguish between conjunction and disjunction on the one hand, and universal and existential quantification on the other hand, in terms of which player has a choice. In the cases of conjunction and universal quantification, the challenger chooses which formula he asks for. Conversely, in the cases of disjunction and existential quantification, the defender is the one who can choose between various formulas. Notice that there is no defence in the particle rule for negation.

Particle rules provide an abstract description of how the game can proceed locally: they specify the way a formula can be challenged and defended according to its main logical constant. In this way we say that these rules govern the local level of meaning. Strictly speaking, the expressions occurring in the table above are not actual moves because they feature formula schemata and the players are not specified. Moreover, these rules are indifferent to any particular situations that might occur during the game. For these reasons we say that the description provided by the particle rules is abstract.

Since the players' identities are not specified in these rules, we say that particle rules are symmetric: that is, the rules are the same for the two players. The fact that the local meaning is symmetric (in this sense) is one of the biggest strengths of the dialogical approach to meaning. In particular it is the reason why the dialogical approach is immune to a wide range of trivializing connectives such as Prior's *tonk*.[19]

The expressions occurring in particle rules are all move schematas. The words "challenge" and "defence" are convenient to name certain moves according to their relationship with other moves. Such relationships can be precisely defined in the following way. Let Σ be a sequence of moves. The function p_Σ assigns a position to each move in Σ, starting with 0. The function F_Σ assigns a pair $[m, Z]$ to certain moves N in Σ, where m denotes a position smaller than $p_\Sigma(N)$ and Z is either C or D, standing respectively for "challenge" and "defence". That is, the function F_Σ keeps track of the relations of challenge and defence as they are given by the particle rules. Consider for example the following sequence Σ:

$$\mathbf{P} - \varphi \wedge \psi, \mathbf{P} - \chi \vee \psi, \mathbf{O} - ?[\varphi], P - \varphi$$

In this sequence we have for example $p_\Sigma(\mathbf{P} - \chi \vee \psi) = 1$.

A *play* (or dialogue) is a legal sequence of moves, i.e., a sequence of moves which observes the game rules. The rules of the second kind that we mentioned, the structural rules, give the precise conditions under which a given sentence is a play. The *dialogical game* for φ, written $\mathsf{D}(\varphi)$, is the set of all plays with φ as the thesis (see the Starting rule below). The structural rules are the following:

SR0 (Starting rule) Let φ be a complex formula of L. For very $\pi \in \mathsf{D}(\varphi)$ we have:

- $p_\pi(\mathbf{P} - \varphi) = 0$,

- $p_\pi(\mathbf{O} - n := i) = 1$,

- $p_\pi(\mathbf{P} - m := j) = 2$.

In other words, any play π in $\mathsf{D}(\varphi)$ starts with $\mathbf{P} - \varphi$. We call φ the *thesis* of the play and of the dialogical game. After that, the Opponent and the Proponent successively choose a positive integer called *repetition*

---

[19] See Rahman, Clerbout, and Keiff (2009).

rank. The role of these integers is to ensure that every play ends after finitely many moves, in a way specified by the next structural rule.

SR1 (Classical game-playing rule)

- Let $\pi \in \mathbf{D}(\varphi)$. For every M in π with $p_\pi(M) > 2$ we have $F_\pi(M) = [m', Z]$ with $m' < p_\pi(M)$ and $Z \in \{C, D\}$.

- Let r be the repetition rank of player **X** and $\pi \in \mathbf{D}(\varphi)$ such that:

  · the last member of π is a **Y** move,

  · M_0 is a **Y** move of position m_0 in π,

  · $M_1, \ldots, M_n$ are **X** moves in π such that $F_\pi(M_n) = \ldots = F_\pi(M_n) = [m_0, Z]$.

  Consider the sequence[20] $\pi' = \pi * N$ where N is an **X** move such that $F_{\pi'}(N) = [m_0, Z]$. We have $\pi' \in \mathbf{D}(\varphi)$ only if $n < r$.

The first part of the rule states that every move after the choice of repetition ranks is either a challenge or a defence. The second part ensures finiteness of plays by setting the player's repetition rank as the maximum number of times he can challenge or defend against a given move of the other player.

SR2 (Formal rule) Let ψ be an atomic formula, N be the move $\mathbf{P}-\psi$ and M the move $\mathbf{O}-\psi$. A sequence π of moves is a play only if we have: if $N \in \pi$ then $M \in \pi$ and $p_\pi(M) < p_\pi(N)$.

That is, the Proponent can play an atomic formula only if the Opponent played it previously. The formal rule is one of the characteristic features of the dialogical approach: other game-based approaches do not have it.

A play is called *terminal* when it cannot be extended by further moves in compliance with the rules. We say it is **X** terminal when the last move in the play is an **X** move.

SR3 (Winning rule) Player **X** wins the play π only if it is **X** terminal.

Consider for example the following sequences of moves:

$$\mathbf{P} - Qa \wedge Qb, \mathbf{O} - n := 1, \mathbf{P} - m := 6, \mathbf{O}-?[Qa], \mathbf{P} - Qa$$

$$\mathbf{P} - Qa \to Qa, \mathbf{O} - n := 1, \mathbf{P} - m := 12, \mathbf{O} - Qa, \mathbf{P} - Qa$$

---

[20] We use $\pi * N$ to denote the sequence obtained by adding move N to play π.

The first one is not a play because it contravenes the Formal rule: with his last move, the Proponent plays an atomic sentence although the Opponent did not play it beforehand. By contrast, the second sequence is a play in $\mathsf{D}(\mathbf{P} - Qa \to Qa)$. We often use a convenient table notation for plays. For example, we can write the play above as follows:

	O			**P**	
				$Qa \to Qa$	0
1	n := 1			m := 12	2
3	Qa	(0)		Qa	4

The numbers in the external columns are the positions of the moves in the play. When a move is a challenge, the position of the challenged move is indicated in the internal columns, as with move 3 in this example. Notice that such tables carry the information given by the functions p and F in addition to represent the play itself.

However, when we want to consider several plays together such tables are not that perspicuous. So we do not use them to deal with dialogical games for which we prefer another perspective. The extensive form of the dialogical game $\mathsf{D}(\varphi)$ is simply the tree representation of it, also often called the game-tree. More precisely, the extensive form E_φ of $\mathsf{D}(\varphi)$ is the tree (T, l, S) such that:

1. Every node t in T is labelled with a move occurring in $\mathsf{D}(\varphi)$,

2. $l : T \to N$,

3. $S \subseteq T^2$ with:

    - There is a unique t_0 (the root) in T such that $l(t_0) = 0$, and t_0 is labelled with the thesis of the game.
    - For every $t \neq t_0$ there is a unique t' such that $t'St$.
    - For every t and t' in T, if tSt' then $l(t') = l(t) + 1$.
    - Give a play π in $\mathsf{D}(\varphi)$ such that $p_\pi(M') = p_\pi(M) + 1$ and t, t' respectively labelled with M and M', then tSt'.

Many metalogical results concerning dialogical games are obtained by considering them by leaving the level of rules and plays and moving to the level of strategies. Among these results, significant ones are given in terms of the existence of winning strategies for a player. We now define these notions and give examples of results.

A *strategy* for Player **X** in $\mathsf{D}(\varphi)$ is a function which assigns an **X** move M to every non terminal play π with a **Y** move as last member such that extending π with M results in a play. An **X** strategy is winning if playing according to it leads to **X**'s victory no matter how **Y** plays.

A strategy can be considered from the viewpoint of extensive forms: the extensive form of an **X** strategy σ in $\mathsf{D}(\varphi)$ is the tree-fragment $\mathsf{E}_{\varphi,\sigma} = (T_\sigma, l_\sigma, S_\sigma)$ of E_φ such that:

1. The root of $\mathsf{E}_{\varphi,\sigma}$ is the root of E_φ.

2. Given a node t in E_φ labelled with an **X** move, we have that $tS_\sigma t'$ whenever tSt'.

3. Given a node t in E_φ labelled with an **Y** move, and with at least one t' such that tSt', then there is a unique $\sigma(t)$ in T_σ where $tS_\sigma \sigma(t)$ and $\sigma(t)$ is labelled with the **X** move prescribed by σ.

Here are some examples of results which pertain to the level of strategies.[21]

Winning P strategies and leaves. Let w be a winning **P** strategy in $\mathsf{D}(\varphi)$. Then every leaf in $E_{\varphi,w}$ is labelled with a **P** signed atomic sentence.

Determinacy. There is a winning **X** strategy in $\mathsf{D}(\varphi)$ if and only if there is no winning Y strategy in $\mathsf{D}(\varphi)$.

Soundness/Completeness of Tableaux. Consider both first-order tableaux and first-order dialogical games. There is a tableau proof for φ if and only if there is a winning **P** strategy in $\mathsf{D}(\varphi)$.

By soundness and completeness of the tableau method with respect to model-theoretical semantics, it follows that existence of a winning **P** strategy coincides with validity: *There is a winning **P** strategy in $\mathsf{D}(\varphi)$ if and only if φ is valid.*

Examples of extensive forms

Extensive forms of dialogical games and of strategies are infinitely generated trees (trees with infinitely many branches). Thus it is not possible to actually write them down. But an illustration remains helpful, so we add Figures 1 and 2 hereafter.

Figure 1 partially represents the extensive form of the dialogical game for the formula $\forall x(Qx \to Qx)$. Every play in this game is represented as a branch in the extensive form: we have given an example with the

---

[21] These results are proven, together with others, in Clerbout (2013).

leftmost branch which represents one of the simplest and shortest plays in the game. The root of the extensive form is labelled with the thesis. After that, the Opponent has infinitely many possible choices for her repetition rank: this is represented by the root having infinitely many immediate successors in the extensive form. The same goes for the Proponent's repetition rank, and every time a player is to choose an individual constant.

Figure 2 partially represents the extensive form of a strategy for the Proponent in this game. It is a fragment of the tree of Figure 1 where each node labelled with an **O** move has at most one successor. We do not keep track of all the possible choices for **P** any more: every time the Proponent has a choice in the game, the strategy selects exactly one of the possible moves. But since all the possible ways for the Opponent to play must be taken into account by a strategy, the other ramifications are kept. In our example, the strategy prescribes to choose the same repetition rank as the Opponent. Of course there are infinitely many other strategies available for **P**.

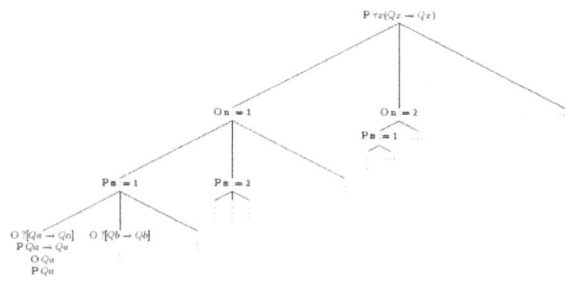

Figure 1.

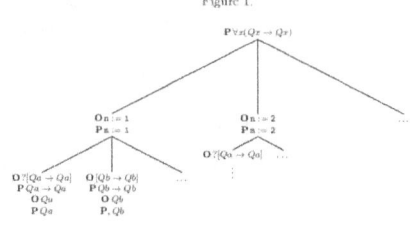

Figure 2.

Appendix 2: Definitional Equality and the Equality-Predicate

Posit	Challenge	Defence
X ! A : set	**Y** $?_{set}$-refl	**X** ! $A = A$: set
X ! $A = B$: set	**Y** $?_B$-symm	**X** ! $B = A$: set
X ! $A = B$: set **X** ! $B = C$: set	**Y** $?_A$-trans	**X** ! $A = C$: set

Table 6.1: Reflexivity, Symmetry and Transitivity within **set**

Posit	Challenge	Defence
X ! $a : A$	**Y** $?_a$-refl	**X** ! $a = a : A$
X ! $a = b : A$	**Y** $?_b$-symm	**X** ! $b = a : A$
X ! $a = b : A$ **X** ! $b = c : A$	**Y** $?_a$-trans	**X** ! $a = c : A$

Table 6.2: Reflexivity, Symmetry and Transitivity within **A**

Posit	Challenge	Defence
X ! $A = B$: set	**Y** $?_{ext}$ $a : A$	**X** ! $a : B$
X ! $A = B$: set	**Y** $?_{d-ext}$ $a = b : A$	**X** ! $a = b : B$

Table 6.3: Simple and Double extensionality

Posit	Challenge	Defence
X ! $B(x)$: set $(x : A)$	**Y** $?_{B(x)-subst}$ $a = c : A$	**X** ! $B(a) = B(c)$: set
X ! $b(x) : B(x)(x : A)$	**Y** $?_{b(x)-subst}$ $a = c : A$	**X** ! $b(a) = b(c) : B(a)$

Table 6.4: Substitution $B(x)$ and $b(x)$

Posit	Challenge	Defence
X ! $I(A, a, b)$: set	**Y** $?_{F-I1}$ **Y** $?_{F-I2}$ **Y** $?_{F-I3}$ [The challenger has the choice]	**X** ! A : set **X** ! $a : A$ **X** ! $b : A$

Table 6.5: Formation of the Equality Predicate

Posit	Challenge	Defence
X ! $a = b : A$	**Y** $?_{def-pred}\, a : A$	**X** ! $p : I(A, a, b)$

Table 6.6: From definitional equality to the equality-predicate

Posit	Challenge	Defence
X ! $p : I(A, a, b)$		
X ! $a : A$		
X ! $q : B(a)$	**Y** $?_{pred-subst}\, b : A$	$q : B(b)$

Table 6.7: Substitution for the equality-predicate

References

Blass, A. (1992). "A game semantics for linear logic". In: *Annals of Pure and Applied Logic* 56, pp. 183–220.

Clerbout, N. (2013). "Etude sur quelques sémantiques dialogiques". PhD thesis. Universities of Lille 3 and Leiden.

Ginzburg, J. (2011). *The interactive stance*. Oxford: Oxford University Press.

Granström, J. (2011). *Treatise on Intuitionistic Type Theory*. Dordrecht: Springer.

Keiff, L. (2007). "Le pluralisme dialogique". PhD thesis. University of Lille 3.

— (2009). "Dialogical logic". In: *Stanford Encyclopedia of Philosophy*. Ed. by E. Zalta. Summer 2011. Stanford.

Martin-Löf, P. (1984). *Intuitionistic Type Theory. Notes by Giovanni Sambin of a series of lectures given in Padua (1980)*. Napoli: Bibliopolis.

Nordström, B., K. Petersson, and J. M. Smith (1990). *Programming in Martin-Löf's Type Theory – An Introduction*. Oxford: Oxford University Press.

Primiero, G. (2008). *Information and Knowledge. A Constructive Type-theoretical Approach*. Dordrecht: Springer.

Rahman, S. (2012). "Constructive Type Theory and the link between Dialogical Logic and Orthosprache. A new start for the Erlanger Konstructivismus?"

Rahman, S. and W. Carnielli (2000). "The Dialogical Approach to Paraconsistency". In: *Synthese* 125 (1-2), pp. 201–232.

Rahman, S. and N. Clerbout (2013). "Constructive Type Theory and the Dialogical Turn. A New Start for the Erlanger Konstruktivismus".

Rahman, S., N. Clerbout, and L. Keiff (2009). "On Dialogues and Natural Deduction". In: *Acts of Knowledge: History, Philosophy and Logic*. Ed. by G. Primiero. London: College Publications, pp. 301–336.

Rahman, S. and L. Keiff (2005). "On how to be a dialogician". In: *Logic, Thought and Action*. Ed. by D. Vanderveken. Dordrecht: Springer, pp. 359–408.

Rahman, S. and J. P. Van Bendegem (2002). "The dialogical dynamics of adaptive paraconsistency". In: *Paraconsistency: The Dialogical way to the inconsistent. Proceedings of the Word Congress Held in São Paulo*. Ed. by M. C. Walter Carnielli and I. D'Ottaviano. New-York: Marcel Dekker, pp. 295–322.

Ranta, A. (1994). *Type Theoretical Grammar*. Oxford: Oxford University Press.

Redmond, J. and M. Fontaine (2011). *How to play dialogues. An introduction to Dialogical Logic*. London: College Publications.

Sundholm, G. (1986). "Proof Theory and Meaning". In: *Handbook of Philosophical Logic. Volume 3*. Ed. by D. Gabbay and F. Guenthner. Dordrecht: Reidel, pp. 471–506.

Thompson, S. (1991). *Type Theory and Functional Programming*. Boston: Addison-Wesley.

Van Bendegem, J. P. (2001). "Paraconsistency and Dialogue Logic. Critical Examination and Further Explorations". In: *Synthese* 127 (1-2), pp. 35–55.

CHAPTER 7

What Can A Sociologist Say About Logic?

Sal Restivo

There is no Logik, only logics (Lotze 1843)

1 Introduction

Suppose that logic transcends society, culture, and history; and suppose furthermore that it transcends space and time. Too many philosophers and theologians are prepared to ignore the fact that in that case we would be speechless and unaware in the face of logic. This is a special case of what I call the apophatic fallacy. Apophatic theology seeks to describe the unknown and unknowable God by negation. Notice that the alternative approach, cataphatic theology, is considered by the apophaticists as limiting God by trying to describe Him positively. In an apophatic framework, a sociologist could have nothing to say about logic just as she could have nothing to say about God. Something like this view of logic has prevailed since ancient times but has not prevented the unfolding of a large literature on the nature, uses, applications and theory of logic. The certainty of Logik, not unlike that of Mathematik and Gottheit, has classically been protected by a triply reinforced iron cage of certainty, authority, and tradition. This iron cage has made

Logik invulnerable to social criticism, skepticism, analysis, reading, and deconstruction. But that iron cage is and always has been an illusion, much more readily penetrated than the warning signs pinned to it are designed to make us think. We have created monsters in math, logic, and God, then forgotten we created them, and as a result they frighten us into silence.

So what do we find when we ignore the warning signs, what do we learn about what logic is and where it comes from? In this essay, I have the modest programmatic goal of exploring what we sociologists might be able to say about logic. I draw heavily on what we have already said about mathematics, and in particular about pure mathematics. I am going to treat logic as an extreme extension of pure mathematics.

In the case of mathematics, I have approached the "comes from" question in terms of the answers offered by mathematicians and philosophers. Everything that has been said about where mathematics comes from can be said a fortiori for logic. It could come from God (*Die ganze Zahl shuf der liebe Gott, alles übrige ist Menschenwerk*, Kronecker declared) or a Platonic realm of Forms; it might *represent* or be one with God. For example, in Mesopotamia, the ratio 2/3 was deified as the god Ea the Creator; and consider that there is a rationale for translating John 1:1 in *The New Testament* this way (Clark 1980): "In the beginning was Logic, and Logic was with God, and Logic was God. ...In Logic was life and the life was the light of men".

What philosophers and other advocates of a transcendental realm of reality fail to come to grips with is that it puts too much of a strain on the scientific imagination to assume that material beings somehow could have access to immaterial realms. Various forms of naturalism, materialism, and critical realism are under the same strain insofar as efforts are made to make them fit into a reality with transcendental, supernatural, or "mysterious" realms and entities.

Perhaps the *Parmenides* is a sign of this strain in Plato. The whole idea of the Forms is left in doubt at the end of Parmenides' critique. Cairns (2005, pp. xviii–xix) concludes that Plato did indeed believe that the "Forms" or "Ideas" exist outside of our minds. But he also suggests the concept may be more earthly than ethereal, in some ways kin to "naturalism, pragmatism, positivism, analysis, and existentialism".

The iconic source of our classical ideas about logic is Aristotle, and even here we find evidence of a latent sociological imagination. He begins Book 1 of the *Posterior Analytics* with the sentence: "All teaching and all intellectual learning comes about from already existing knowledge". He makes it clear that he means this to apply to the mathematical sciences, to "each of the other arts", and indeed to both deductive and inductive arguments (Aristotle 1984, p. 114). The subtlety here is indicated by (Weisheipl 1970): "Instead of saying that all knowledge is actually in the mind or actually not in the mind, Aristotle insists that all knowledge is potentially in the mind and the business of learning is to draw this potentiality into actuality." The important point here is that Aristotle seems to introduce activity in the world as a condition for knowing.

Even if one can find hints of social theory in the Platonic views of mathematics and the Aristotelean view of argument, the image of something "outside" of us —something transcendent, godlike, pure, abstract— keeps mathematics and logic ultimately separated from the social and material realms of experience. For the sociological theorist, references to realms "outside" of us are mistakes in reference; they are in fact to be understood as pointing to social referents. Emile Durkheim and George H. Mead pioneered in the development of sociological theory as the rejection of transcendence, immanence, and psychologism. Consider in particular Dirk Struik's conception of the goal of the sociology of mathematics: "to haul the lofty domains of mathematics from the Olympian heights of pure mind to the common pastures where human beings toil and sweat" (Struik 1986, p. 280).

In spite of the widespread support in mathematics and the philosophy of mathematics for Platonism, supporters have not been able to escape the self-contradiction, and even the absurdity, of the transcendence claim. If mathematics is outside of space and time how can we reach it from our earthbound grounds? It is clear that exactly the same problem arises in the case of God when the faithful argue that God cannot be captured by our eyes, our words, or our minds. What can such claims possibly mean? Any effort to answer such a question will mangle reason, experience, and understanding. This is an expression of the apophatic fallacy applied to logic. Within the material (and more generally "the natural") world there is an often overlooked source for logic, mathematics, science, and God; that is the social world. Here too we find the roots of the very ideas of transcendental and supernatural realms. The only reasonable answer, I argue with my fellow sociological materialists, is that logic comes from the social world, and in particular from social networks.

The idea that mathematics as a vocation is social would not be disputed by most working mathematicians. The trouble begins when the sociologist wants to draw out the more technical meanings of "social". The sociologist wants to press the idea of the "social" beyond its everyday meaning and to argue (1) that there are more dimensions to social interaction than are evident in everyday social life, and (2) that mathematical objects themselves are social. The transcendental realm is a cultural creation, not a reality out of space and time. So is the supernatural, and so then are the gods and God. The fundamental project of the sociological sciences can indeed be viewed as locating the everyday world referents for transcendental, supernatural experiences, and in general experiences that seem mysteriously without earthly referents. The act of "referencing" experiences is dependent on the progress of our knowledge and understanding of the world, the progress of the sciences and more generally of the learning arts.

What can we conclude as social constructionists, that is, as sociological and materialistic critical realists? The idea that mathematics and logic are pure or transcendent is "an expression of the felt autonomy of the inner activities of the intellectual network" (Collins 1998, p. 878). The certainty of mathematics and logic are a function of how tightly the generational links across mathematical networks are interwoven. It is also a function of the continuity across generations. The "chain of social conventions" in mathematics is robustly repeatable. It is this robustness that accounts for the sense of certainty mathematicians and logicians, along with laypeople, share about mathematics and logic.

Neither truth, certainty, nor thought itself "arise in isolated brains or disembodied minds" (Collins 1998, p. 877). They all arise in social networks. Consciousness itself, as Nietzsche already recognized, arises in and is a network of social relations. At the end of the day, sociologists are wont to ask "How could any of these phenomena arise anywhere else, be anything else; what is there that is anywhere else"? It is discourse, with its "objective, obdurate quality", that produces that "strong constraint that answers the concept of truth" (Collins 1998, p. 865). Even the most elementary exercise in mathematics or logic, indeed even the most elementary understanding of an equation, engages us in a form of discourse (and more broadly, in Wittgenstein's terms, a form of life), a network of teachers and students, of researchers, inventors, and discoverers. The "universality" of mathematics and logic, like the universality of any cultural system, trait, or representation is grounded in the universality of its social practice and discourse. "Universality" is a function of the diffusion of discourse by means of social exchange (trade, communication, military and scientific missions).

2 Toward an Archaeology of Logic

As soon as we question (for whatever reasons) the unity, purity, and universality of logic, "it loses its self-evidence; it indicates itself, constructs, only on the basis of a complex field of discourse" (Foucault 1972, p. 23–

24). Foucault, of course, was not thinking of logic or mathematics here. He treated mathematics (and by implication logic) as something of a special case, immune to the power of his archaeological method. He lost his courage here, so let us take on the task and ask some Foucauldian questions about logic.

Logic is an everywhere dense discourse. Do logical symbols hide something? Are they embedded in networks of power, and are they deployed in ways that purposefully obscure the power behind their visual and oral re-presentations? How is it then that logic seems to have escaped matter? How has logic hidden the fact that it is a (indeed *the*) discipline that disciplines? How do we reveal the systems of regularities that determine logicians by determining their situations, functions, perceptions, and practical possibilities? How do we reveal the social, cultural, and historical conditions that "dominate and even overwhelm" logicians? Logic is more than a discourse, more than a language. It is an institution embedded in a culture and assigned a more or less specific domain of control.

We need, following Foucault's method, to be able to reconceptualize the problem of logic not as a problem in ontology (or even in classical epistemology) but as a problem in politics and ethics (or a problem in morals). Let us pursue this Nietzschean turn. Individuals do not make decisions about what is right and wrong or true and false on their own. Such decisions are settled by institutions. We are born into classifications, logical operations, languages, and metaphors. It is on the basis of such considerations that sociologists of knowledge of my type reach the conclusion that logic is a moral order. It is important to keep in mind Durkheim's remarks on the categories of space, time, and causality. These are the most general relations between things, and they dominate our intellectual and everyday lives. Communities of men and women must be in accord about these essentials at any given historico-cultural moment. Without this accord, they would not connect intellectually, emotionally, and linguistically. Humans are not free to choose or deny "the categories". Social life requires a minimum consensus without which society dissolves. This makes adherence to social norms and expectations a moral imperative, a moral necessity. Keep in mind these categories are treated as if they are a priori but they reflect, arise and crystallize in social and cultural contexts. This consensus rule allows for various degrees of deviation but communities break down very quickly if deviations reach even modest levels. The sources of deviation are based on factors such as the movement of material resources, engagements with hostile forces, and disease.

What role do numbers and formal symbols play in grounding our ideas or experiences of abstraction, purity, and the sacred? How do numbers and formal symbols play into constructing and sustaining boundaries and relationships? The moral necessity of logic is enhanced as its professional boundaries are constructed and concretized around those thought communities and thought collectives (Fleck 1979 (1935)) dedicated to these very ideas. Logic develops a capacity to overtake and dominate all other forms of reason by persuading leading minds that it is the God of reason. It does this by becoming purity, abstraction, and generalization in extremis.

All institutions provide the categories of thought, set the terms for knowledge and self-knowledge, and fix identities. But more than this, they "must secure the social edifices by sacralizing the principles of justice" (Douglas 1986, p. 112). In logic, classifications and theorems, proofs and conjectures are held together by the sacred glues of logic itself and logicized canons of reason. Given the sociological conception of the nature and function of institutions, it should not be surprising to find that questions and issues of morals merge with questions and issues of what is real and what is illusory, what is true and what is false.

With Spengler, I have claimed that: "*Es gibt keine Mathematik*, es gibt nur Mathematiken". That remark can be considered a key moment in the emergence of a sociology of mathematics. Now, echoing Lotze (1843), I make the parallel claim for logic: "Es gibt nicht die Logik, es gibt logische". This then puts us on the pathway to a sociology of logic.

Purism and some sort of technicism have a mutual affinity. The purer, more formal, and more mechanical a discourse is, the easier it is to claim that it can be used to break down common language barriers. Pure mathematics, for example, has been described as a notational doctrine for relatively ordered thought operations which have been mechanized. Proofs, perhaps the central apparatus of pure mathematical work, indeed seem to be machines for factoring out human agency. They are an important part of the material validation of theorems, devices for transforming theorems into matters of fact.

Shapin and Schaffer (1985) have shown that in experimental settings in the physical sciences, *machines* help to eliminate human agency in establishing facticity. Thus, Boyle, in criticizing one of Huygens' experiments, raises questions not about the experimenter's "ratiocination" but rather "the staunchness of his pump". In logic, a theorem is an experimental result. The experiment in this case is none the less an *experiment* (and social) for being an immediate product of mental effort (at least apparently; see Amann and Knorr Cetina (1989) "thinking through talk"), nor any less an experiment for being in part a product of more or less unconscious thinking.

The proof-machine is offered as a material, objective validation of the logician's work. In effect, the logician says: "It's the proof that says this, not me". The testing and refinement, the acceptance and rejection of proof brings other logicians into the picture to bear "collective witness" and to make the operation of the proof a collective performance. In the end, if the proof is accepted, it is a result of the public constitution and validation of knowledge, under the dictum: "it is not I who says this, but *all* of us". This constitutes the social validation of a theorem as a matter of fact. When the validation occurs in textual contexts through "virtual witnessing", literary resources help to factor out human agency and to construct a community of consensus. It should be noted that establishing a matter of fact does not preclude later challenges and reversals. And proofs, like other technologies in general, can become obsolete. Abel's 1824 proof that equations of higher degree than four cannot be solved by root extractions except for special values of the coefficients is an example of an obsolete proof. It is, furthermore, interesting to note that the "inevitability" of the development of logic is widely considered to be a feature of technological development. (Edge and Mulkay (1976), for example, point out that many of the astronomers they interviewed held this view of technical developments in radio astronomy). If we recognize that logic is a technology, it becomes easier to understand what appears to be an "inner logic of development" as a matter of social practice, and to see that the process is neither mysterious nor mystical.

The final stage of the purification of reason is its transformation into logic. And mathematics, to the extent that it develops independently, is eventually overtaken and taken over by logic. At that point, logic not only leaves mathematics behind but the world. What have sociologists made of this phenomenon?

3 The Sociology of Logic

We have had a sociology of logic developing within the sociology of mathematics since the pioneering interventions of Bloor, Restivo, and MacKenzie in the 1970s and 1980s. Not only that, we have on hand two major empirical investigations of logic as practice—one in the ethnomethodological tradition by Eric Livingston and one that rests somewhat more firmly in the mainstream of sociological thinking by Rosental (2008). My own sociological reading of Boole's *Laws of Thought* (Boole 2009 (1854)) and Kleene's *Introduction to Metamathematics* (Kleene 2009 (1950)) is relevant here, as is Andrea Nye's feminist reading of the history of logic (Nye 1990), all considered below.

Consider, to begin with, Eric Livingston's ethnomethodological approach to logic, a sociology of logic in practice but one that is limited in its sociological import. The limitations follow from the sociologically idiosyncratic methods and assumptions of ethnomethodology.

Livingston addressed two core concerns in the philosophy and sociology of mathematics: what are mathematical objects, and what is the source of the compulsion associated with mathematical reasoning. The answer to these questions lies in attending to the moment-to-moment work of the mathematical reasoner(s) at the site of the mathematical work. We see in this attention to the "living foundations of mathematics" what looks like the ethnographic approach to scientific practice pioneered by the new sociologists of science in the late 1960s and early 1970s. Livingston's work is designed to reveal the social processes behind logic as understood in the classical tradition, that is, logic as a pure, abstract manipulation of formal symbols. The major focus of Livingston's work is Gödel's theorem and Livingston clearly has a firm grasp on the technicalities of the theorem. Livingston's approach is to walk us through the proofs in Gödel and in the simpler case of Euclid. What, in other words, is going on in the unfolding of a proof?

Ethnomethodology does not pretend to theoretical explanation, does not debunk or demystify, and does not promote understanding by way of historical and scientific analyses. Without denying the innovative and informative nature of this study, sociologists like myself are left to wonder what exactly has Livingston achieved here? My answer is one that I argue applies to the ethnomethodological approach in general; what it achieves is a translation not an explanation (Rosental 2008 is kinder to this tradition and rejects the translation viewpoint).

From the standpoint of the ethnomethodologist this is not a devastating criticism. They achieve what they set out to achieve, and that is to give some sort of account of what it is like to do the kind of work they are studying in any given case. But the failure to offer an explanatory account —the goal of any scientific inquiry— makes their project one that fails to further our understanding of the phenomenon at issue. Bloor's criticism of Livingston's approach comes to the same thing—Livingston fails because he refuses to theorize, he refuses to "do science" (Bloor 1987, p. 351). And he fails to theorize because the ethnomethodologist is obliged to make the distance between him/herself and his/her subject as small as possible. What this means is that in the case at hand, Livingston does more mathematics than he does sociology (or philosophy). Bloor shows clearly that there is an unarticulated theory here, a locality theory. We ethnographers of science expect to find science in the contexts, contingencies, and conditions of the work environment. This is where Livingston leads us, but he doesn't end up in an environment of people working together but rather in an abstract "primordial setting", a system of pure mathematical work in a closed system apart from history, culture, and profession. I will leave the Livingston story here with a recommendation that the interested reader consult Bloor's outstanding review for a full analysis of what Livingston has accomplished and what he has failed to accomplish. For the sociologist of logic in my or Bloor's sense, Livingston provides some interesting and useful data but does nothing to further our explanatory agenda.

Consider next Boole's *Laws of Thought* (1854) which was to ground a science of the mind in observations. This was a different endeavor than studying the external world of nature. The laws of Nature, according to Boole, are not in general accessible to immediate perception. Certainty may be ever more closely approached, but it is never achieved. By contrast, knowledge of the laws of the mind appears in "particular instances". The truth of such laws requires no repetition of confirming instances, and no large set of observations.

Boole fails the "sociological awareness test" by not recognizing that Aristotle's dictum *de omni et nullo* (whatever is affirmed or denied of a whole may be affirmed or denied (respectively) of any part of a whole; all valid syllogisms are reducible to applications of *dictum de omni* and *dictum de nullo*), and the so-called "categorical propositions" (e.g., All Ys are Xs) are in fact high level exercises of generalization grounded ultimately in inductions based on experience. The repetition over great expanses of time of such experiences and their institutionalization in "common sense" is the source of the sense that certain things are "self-evident".

The flaw in Boole's thinking arises from his failure to see himself as a product and agent of culture—or, more radically, as a vehicle for cultural experiences he has internalized and experienced as a capacity to clearly apprehend a single instance of an event or phenomenon (thus subjecting himself to the fallacy of introspective transparency).

If Boole fails to see himself as a vehicle of culture, it is not because of any resistance to seeing himself as a vehicle. Science, Boole says, is the business of discovering laws, not creating them. Our minds are not our own; we do not constitute them and our intellect is not the product of our will. Science, then is not dependent on individual choice. In his pursuit of the laws of thought, Boole is guided above all by his own sense of self-evidence. Formal laws are based on observations and reflections, Boole writes. But "results" are independent of whether we treat theory as grounded in experience or as a matter of strict deduction. It is notable that Boole wanted his work to gratify the intellect but also to contribute to "human welfare".

From Boole's perspective, science gives us primary (fundamental) and secondary (derived) truths. Boole focuses on the fundamental truths, laws and principles from which all the rest of science may be deduced and into which all may be again resolved. The test of the "completeness" and "fundamental character" of the laws of science is the completeness of derived truths and the general methods used in science.

Boole believed that commonalities and universals across human cultures reflected the laws of thought. But of course the situation is that genetic and biological commonalities interact with "external" (including social) commonalities to produce mental commonalities. Boole mistakenly assigns priority to the "awakened" (my term) mind, the socialized mind, without acknowledging the social foundations of thought. He is thus left with no alternative but to argue *from* universal laws of thought rather than *to* socially constructed categories of thought. The mind is not the pristine, a priori genetic-biological instrument Boole thinks it is.

Boole gives us an excellent opportunity to watch the progress of moving up levels of generalization ("abstraction") from the "primitive" ground (or frame; see Goffman 1974) of everyday life. He shows that the symbols of Logic he introduces are subject to "the special law" $x^2 = x$. But having introduced this formal law, he goes on to indicate its "primitive" roots (Boole 2009 (1854), p. 37). What in fact Boole does is move into a fantasy world in which there are only two material resources, the numbers 0 and 1. Like things in our everyday world they are subject to certain lawful relationships, in this case $0^2 = 0$ and $1^2 = 1$; thus $x^2 = xx$, considered algebraically, has no other roots than 1 and 0. Only differences of interpretation will apply within this Algebra. Thus does Boole set forth the principle upon which his method rests. However, if we look at the earlier paragraphs leading up to Boole's laws, we notice that $x^2 = x$ is developed in a materially grounded way. First he shows that $xy = xy$ in his developing system is based on a "class" perspective and grounded in examples such as "white things" (x), "sheep" (y), and "white sheep" (xy). He then argues that the combination of two literal symbols in the form xy expresses the whole of that class of objects to which the names or qualities represented by x and y are together applicable. It follows that if the two symbols have the same signification, their combination expresses no more than either of the symbols taken above would do. This leads to $xy = x$, and then (since y has the same meaning as x), to $xx = x$. Finally, by adopting the notation of common Algebra, Boole arrives at $x^2 = x$. We are now back in the realm of his earlier use of $1^2 = 1$ to represent the phrase "good, good men". So Boole constructs a "primitive" everyday world in which only 0s and 1s exist. Eventually Boole gives 0 and 1 in Logic the respective interpretations

Nothing and *Universe*.

The mind, Boole claims, arrives at the existence of a universe as a deduction from experience, or *hypothetically*. Either way we are dealing with a social mind, a socialized mind. Boole's implicit recognition of this occurs at the end of a critical discussion of the syllogism. The syllogism is associated with the development of language, a social process. Intellectual processes involve memory, and usage; and certain canons of ancient logic have become inculcated in the very fabric of thought characteristic of a cultured mind. This has to apply to Boole. He has inculcated the "texture of thought" characteristic of his time and place.

It is commonplace in fields that are considered "pure" to find workers oriented to "unity and harmony". Even if other values such as power and efficiency are acknowledged, they are subordinated to values of unity, harmony, fitness, and beauty. This conception would be most fully realized if even the very forms of the method were suggestive of the fundamental principles, and if possible of the *one* fundamental principle, upon which they are founded. I suggested earlier a connection between pure disciplines and religious or theological quests. The connection between such quests and the isolating effects of professionalization and specialization cannot be considered further here. That we are dealing, in any case, with a world view nourished by notions of gods and kings ruling orderly domains, and especially of an omniscient, omnipresent God-King is suggested by Boole's commitment to identifying a central pervading law. This is not a "mere metaphor"; on the other hand, it is not necessarily unrelated to developing a capacity for gaining personal or collective control over some area(s) of the worlds of self, society, and nature. We must take seriously Boole's reference to the "Author of Nature" and His "immutable constancy" as an indicator of what it is about the world that is significant for Boole. Boole's logic is in fact part of a strategy for establishing the existence of God and Universal Morality. Logic, like pure mathematics and God, can serve as a strategy for gaining control over a world that threatens momentarily to reveal itself as a world of "chance and inexorable fate". If the order in a person's or community's life is felt to be fragile and in need of an anchor for its security, then there will be a search for —and even the fabrication of— order; thus the coordinated search for God, Beauty, Truth, Logic, and Purity among pure mathematicians, logicians, linguists, and artists. All the better if the source of our moral rules and guidelines can be located outside of ourselves in the most extreme way by placing the source in a transcendental realm.

Given that Boole can be situated between two generations of mathematicians concerned with issues of Logic, and given that there are readily identified organizational changes across these three generations in the directions of specialization and professionalization, then we should find a decreasing emphasis on common language and (in the case considered here, secondary propositions) a decreasing emphasis on time. The orientation to abandoning ties to the everyday world can be related to organizational changes that induce a timeless view of individual and collective life. Professionalization, a process that was already operating in the mathematical community of Boole's time, is a key determinant of the orientation to time. The process of professionalization fosters the abandonment of everyday time. First, it removes the professional from the time frame of the everyday world. It creates a new frame for professional time. Thus, a certain *kind* of time is abandoned. But the new time is more flexible, perhaps more general (or as some would say, more abstract). Time then becomes (like God) remote for the modern secular universalistic professional. This notion is compatible with a mundane time perspective that dictates short term activities and which professionalization may actually intensify. In the extreme case, universalistic standards and eternalist orientations to such notions as "making a contribution", receiving eponymous rewards and constraints (working "outside time and space") may make time disappear.

How does Boole arrive at the idea of "the perfect liberty which we possess" when it comes to choosing and ordering while pursuing the implications of premises in given demonstrations of propositions, that is, when it comes to "determining what elementary propositions are true or false, and what are true and false under given restrictions, or in given combinations"? This idea of liberty is not grounded in some sort of organic sense of free will but rather in a mechanistic framework (Boole 2009 (1854), p. 185). Rigorously demonstrating real premises is achieved when we remove all doubts and ambiguities. Boole's objective, realized in arranging the order of premises and demonstrating their connections, is carried out with indifference. Inference is conducted precisely and mechanically. The inference machine Boole aspires to invent is a perfect companion for his purism, and is analogous (in part if not entirely) to the mechanical social rituals designed to guarantee or underwrite truths. Boole is on the threshold of the purification of modern mathematics. I want to jump ahead now to the relatively recent past and look at metamathematics. The general strategy I have outlined in this chapter can help to make sense out of highly generalized ("abstract") mathematical work without recourse to non-materialist (and especially mentalistic or cognitive) categories and "explanations". The case I examine shows how a given set of generalizations ("abstractions") can become the raw materials of later everyday work. When this occurs (and this is characteristic of science-oriented fields and in fact all fields oriented to generalizing principles), the reaction among workers will tend to get grounded in a philosophy of naive realism. This then becomes the basis for operating on old and creating new generalizations. The greater the extent to which their work is removed from the context of everyday (mundane) work, the more difficult it is for them to gain access to the social and material groundings of their work.

S.C. Kleene's *Introduction to Metamathematics* (Kleene 2009 (1950)) is the focus of my discussion here for two reasons. It was for a long time a leading textbook introduction to metamathematics by a leading practitioner; and —more importantly— Kleene's exposition is detailed and clear, and makes it relatively easy to identify the roots of pure ideas in material reality and the cultural continuities that make "pure" work possible. In a way, Kleene makes the sociological and materialist case all by himself. The relevance of his work to the sociology of logic lies in the way he allows us to examine the process of constructing higher and higher levels of generalization. This process in mathematics leads inevitably to the development of or convergence with logic. I should note that the development of linguistics plays a role in the unfolding of the Parmenidean goal. We see this connection in the works of Leibniz, Hamilton, and others.

Kleene (op.cit., p. 59) states that propositions embody the results of mathematical work. A mathematical theory is constructed out of propositions, that is, it is a set or system of propositions. Propositions and systems of propositions are objects in mathematical reality. Now we notice that the continuum of real numbers provides the fundamental system of objects for analysis (op.cit., p. 30). In the arithmetization of analysis, real numbers are defined as certain objects constructed out of natural numbers, integers or rational numbers. Kleene writes: "... in the arithmetization of analysis, an infinite collection (of rationals forming the lower half of a Dedekind cut, or of digits in sequence forming a non-terminating decimal, etc.) is constituted an object, and the set of all such objects is considered a new collection. From this it is a natural step to Cantor's general set theory".

Notice that mathematical objects such as integers are constructed in relationship to non-mathematical objects such as cows, apples, fingers, and so on. Kleene refers to a later development in mathematics associated with the transformation of mathematical work into a highly specialized activity where mathematical objects become the materials out of which new mathematical objects are formed. The referent mathematical objects can be the source of models, that is, they can be taken as things in mathematical reality that are analogous to cows, apples, etc.

or they can be directly manipulated and used in a sort of tinker-toy way to create new objects. So they can be used as sources of generalization or as material resources. There is a kinship of sorts here with the two ways of introducing systems of objects into mathematics identified by Kleene (op.cit., p. 26–28). The *genetic* or *constructive* method is illustrated by the manner in which the natural numbers are generated. Kleene has in mind the inductive definition of natural numbers. But this in nothing more or less than the development of the natural number ideas in our commerce in the natural world. In the *axiomatic* or *postulational* method, we begin with some propositions that are assumptions or conditions on a system of mathematical objects: The consequences of the actions are then developed as a theory about any existing system S of objects which satisfy the axioms.

In the formalization process, what happens in effect is that the mathematician explicitly creates a mathematical reality very much the way a Tolkien or a Frank Herbert creates a fantasy world. And the mathematician, like the science fiction or fantasy writer, carries over into his or her new world certain preferences and notions, and indeed a worldview. In the case of the mathematician, a philosophical position or world view is carried over that is analogous to the view of the naive realist natural scientists. Thus Kleene (op.cit., p. 62) writes: "the object theory is described and studied as a system of symbols and of objects built up out of symbols. The symbols are regarded as various kinds of recognizable objects".

Note that the metatheory, the theory about the object theory, is intuitive and informal, and expressed in ordinary language using mathematical symbols (op.cit., p. 62): "the assertions of the metatheory must be understood. The deductions must carry conviction. They must proceed by intuitive inferences, and not, as the deductions in the formal theory, by applications of stated rules. Rules have been stated to formalize the object theory, but now we must understand without rules how these rules work. An intuitive mathematics is necessary even to define the formal mathematics".

It is clear from the way Kleene keeps introducing the need for intuitive mathematics that we must attack the naive realism of formal mathematics in quite the same way that we attack naive realism in the sciences in general. Human agents (mathematicians or logicians here) create a world of objects; a *culture* then applies itself (that is, a network of mathematicians with shared values, paradigms, worldviews) moves into that world and goes to work. The object is to understand and explain those objects. The mathematician or logician creates a world, then gets "born" into and raised in it as a member of a culture; this is a world of chaos for him or her. There are no immediately known, a priori rules. The mathematician now sets about unraveling the nature of that world. This is something like a God coming down to earth to study *His* creation. This social reproduction of the world view of objective science qua naive realism is explicitly illustrated in Kleene's (op.cit., p. 63) claim that: "Metamathematics must study the formal system as a system of symbols, etc. which are considered wholly objectively. This means simply that those symbols, etc. are themselves the ultimate objects, and are not being used to refer to something other than themselves. The metamathematician looks at them, not though and beyond them; thus they are objects without interpretation or meaning".

The rationale for a constructivist sociological interpretation here is provided in great part by Kleene himself, even while he adheres to a form of naive realism. What we see is the self-conscious creation of an object world (world II) by a vehicle for a thought collective, and a product of a natural and social object world (world I). Object world I encompasses object world II and its product, the metamathematician; the whole process stylizes the idea of objective science. And we see how, just as in the case of social construction of the gods, people can come to alienate themselves from the things that they themselves have manufactured. The laws of the object world (world II) can be called world III. Indeed, there is a correspondence here to Popper's 3-worlds philosophy. Popper's world 3 is ambiguously pinned to world 1 and world 2 and can easily take off into a Platonic realm of Forms.

The idea of "operation" is crucial for understanding mathematical activity as a social, material process. Let us begin (with Kleene 2009 (1950), p. 125–126) by noticing that in the elementary school arithmetic of positive integers, the numerals 1, 2, 3, ... had meaning in terms of counting and measurement. But when it came to the addition and multiplication tables, those numerals could be "any enumeration of distinct objects". From this standpoint, the arithmetic deals with operations, i.e., functions, + and × over a domain of objects (1, 2, 3, ...), and depends only on the possibility of recognizing and distinguishing between those objects, and not on their intrinsic nature.

Kleene now follows the same procedure to set up a new arithmetic. He constructs a domain of two objects and four operations or functions. In effect, he creates *six* objects, since the four functions are, metamathematically speaking, "meaningless given objects". The two objects to be operated on are t and f; the four functions are: $\wedge$, &, $\vee$, $-$. It is important to note that these "meaningless" objects are all part of the cultural baggage of mathematics. Let me let Kleene explain what he is up to; and note that this is yet another example of operating on operations reintroduced as objects (op.cit., p. 126): "we introduce a metamathematical computation process (called a *valuation procedure*), by which a function in the arithmetic (or a table for such a function (called a *truth table*) is correlated to each of the symbols ... and thence to each proposition letter formula. Then we study metamathematical properties of proposition letter formulas defined in terms of correlated functions (or tables)".

Note that Kleene chooses "suggestive" symbols (in this case, t and f suggest the notions of "true" and "false" in the logical interpretation), even though it is theoretically immaterial what symbols we choose—so long, of course, as they can be distinguished from one another. It is common in mathematics to employ "the same designation for analogous notions arising in related technical theories" (op.cit., p. 139). Gödel's famous results follow this pattern of getting into the structure of a formal system as a system of objects. The mathe-logical world that Gödel enters is, of course, a world constructed by Russell and Whitehead, *Principia Mathematica*. He enters this world's structure to great depths. Kleene (2009 (1950), pp. 205, 246) describes what this process is like in part: "...by selecting a particular enumeration of the formal objects, or a particular correlation of distinct natural numbers to the distinct formal objects (not using every number), and then talking about the correlated numbers instead of formal objects, metamathematics becomes a branch of the arithmetic of the natural numbers [Gödel numbering]". For a more detailed sociological study of these two cases, Boole and Kleene, see Restivo (1992, pp. 149–176).

The most important contribution to the empirical sociology of logic is Claude Rosental's *Weaving Self Evidence* (Rosental 2008). Rosental engaged the realm of logic by wondering if it was possible to grasp sociologically the process of producing a logical theorem. All of his expectations had already been realized in the ethnography of science beginning in the early 1970s. What is original about Rosental's research strategy is not the strategy itself but applying that strategy to an extreme hard case in the sociology of knowledge. The strategy he adopted involved locating his study at the intersection of technologies of proof and forms of ostentation. Ethnographers of science will not be surprised that Rosental found a diversity of practices mobilizing heterogeneous resources at the center of the production of logic.

We learn here, as we learned from the pioneering ethnographies of science, that writing is at the center of the development of "economies of conviction". When we attend to social practices in the sciences, mathematics, and logic we see material practices rather than the exchange of "immaterial" ideas predicted by classical paradigms in the history and philosophy of science. Logic cannot be "reduced" to a process of reasoning.

Let us pause a moment here to consider Rosental's concern with "reductionism". The fact that he considers social constructionism reductionist demonstrates that he has failed to grasp the essence of reductionism and failed equally to grasp social constructionism as the fundamental theorem of sociology. I have been at pains in several papers and lectures to argue for social constructionism as the fundamental theorem of sociology. Unfortunately, issues in the unfolding of the new sociology of science in association with the science studies movement have led to the mistaken assumption widely held throughout the intellectual community (and not exempting sociology itself) that social constructionism entails some form of relativism. Social constructionism is a scientific idea fully compatible with the most advanced notions and forms of realism.

It is important to keep in mind that the new sociology of science, in conjunction with the science studies movement and postmodernism has changed our understanding of the very idea of science and the terms of scientific discourse including truth, objectivity, and realism. Science itself has to be considered in at least two senses. Small "s" science is the fundamental and primordial mode(s) of human reasoning strategies. Capital "S" Science is science as a social institution, a social practice embedded in the modern modes of techno-industrial production (loosely, the knowledge system, the system of rationality associated with "capitalism"). Realism is no longer the naïve version featured in classical philosophy or the traditional scientific worldview. In both cases, and in the case of the entire lexicon of science, these concepts have for the most part become grounded materially and socioculturally, and all Platonic idealisms have been erased.

It is thus curious to find sociologists of knowledge such as Rosental recognizing on the one hand and correctly that the experience or practice of logic cannot be explained in terms of "reasoning" or interactions between individuals; the focus has to be necessarily on group dynamics. On the other hand, Rosental argues that "rematerializing" the debates among logicians helps us escape social relations in the "strict sense of the term". In fact, if Rosental is worried about avoiding some sort of sociologism, then the solution is already abroad. Social constructionism as I have construed it is intimately embedded in a social ecology and indeed in a sociological materialism that escapes the strict sense of social relations.

All of the factors Rosental identifies as implicated in the sociology of logic are encompassed by the version of social constructionism I speak for. It is instructive to note that Rosental turns to philosophers at critical moments to support his arguments and not to the sociologists of science who would seem to be better placed to support him. Again and again he misses opportunities to acknowledge sociologists who have contributed to empirically establishing the diversity and heterogeneity of the sciences, and the contexts of debate in the sciences and mathematics. He is therefore led to turn to Ian Hacking to defend his understanding of the limits of social constructionism, a philosopher who treats social constructionism as a philosophical idea rather than a scientific concept.

Rosental argues that we must restore eyes and hands to those who produce and manipulate formalisms. To support this argument he turns to a cognitive scientist (Edwin Hutchins) when he could more easily find support among his colleagues in the sociology of knowledge, and more directly relevant support. My point here is not to promote disciplinary imperialism but rather to argue that there is something invidious about turning away from those closest to your training, interests, and research focus and going to relative "outsiders". It diminishes and makes invisible the contributions of sociology and subordinates it to traditional disciplines that tend to ignore, dismiss, or misconstrue sociology. Rosental seems to assume that cognitive scientists and philosophers have more intellectual legitimacy in the world of research than do sociologists.

Rosental's results regarding plurality, the lack of a universal consensus, and the wide variety of elements debated by the logicians are just what we would expect given the history of the new sociology of science and the assumption that logic is a social practice. A closer attention to the sociology of science literature might have led Rosental to see that what he calls "tangential viewpoints" and a "wide variety of debated elements" are just constituents of the facts of logic.

In identifying the work of "de-monstration" as a major finding of his research, Rosental again misses the point that all of the ingredients of de-monstration are what social constructionism is meant to capture. "De-monstration" is used to point to a strong form of demonstration that draws out what is "between the lines" so to speak, that is, to construct a new layer of inscriptions that will literally lead the reader, viewer, or listener to a specific conclusion. The practice here is grounded in the use of eyes and hands —one might perhaps say postures— and not on a mode of reasoning that links thinking minds.

In the end, Rosental finds that logical statements are produced and stabilized in the same diverse, heterogeneous ways as the facts of the experimental sciences. This is not as "unexpected" as Rosental seems to think. After all, we have already found in the sciences and mathematics that facts and findings are constructed in social worlds out of the material and social constituents of those worlds. The more organized these social worlds become, the harder it becomes to ground the productions in that world in the worlds of everyday life; the result is that these productions are projected into a transcendental world. The unfolding of sociology and anthropology has had as one of its key consequences the rejection of the transcendental and the supernatural. The new sociology of science in particular has shown us how to ground apparently transcendental, Platonic ideas in the world of materials and humans. In this regard, Rosental has helped us move further along in this unfolding by underscoring empirically that —as we had already hypothesized and as Durkheim had already theorized— not even logic escapes reality.

In spite of his affinity for Durkheim, Rosental is too enamored of the ethnomethodologists and the Latourian ANTs (actor-network theorists) to see that Durkheim is above all a social constructionist. This incidentally helps explain why Rosental does not agree with those who like me view ethnomethodology as a process of translation. The Durkheimian tradition, in fact, has been instrumental in opening up a road to grasping the dynamics behind the production of certified knowledge claims. Rosental's work has more in common with that tradition as it unfolds in the works of Ludwig Fleck, Mary Douglas, David Bloor, and myself than it does with the ethnomethodologists and ANTs. His failure to align himself more fully with that Durkheimian line is rooted in the fact that he conflates social constructionism and sociologism, and views this conflation as promoting relativism. It appears that it is just these flaws in his perspective that lead him to stress correctly that his findings do not show that logic is "inconsistent" or "irrational". This is just what logic is, this is just what the facticities of logic are. And they are no less factual, no less useful, no less "real" because they have been shown to originate in the spheres of human life rather than the illusionary realms of the transcendental and Platonic. All of our ideas —whether true or false— can be come to only through our social practices and discourses in our social ecologies.

Suppose once again that logic exists outside of time, space, history, society, and culture. It would then be unreadable. There could be no account of logic that was not an account of the truths of logic. Nye's achievements (Nye 1990) are a consequence of her courageous unwillingness to accept the logician's account of "himself" on "his" own terms. Nye assumes that logic, like all human activities, must be motivated by and embody desire. So she sets out on a difficult trek with the expectation that her assumption would lead to a reading of logic. This is in fact the motivation behind all post-1970 sociologies of knowledge. There can be no human activity, no practice, no discourse that transcends time, space, history, society, and culture, none that cannot in principle be read.

Where should we begin our effort to read logic? Aristotle of course is an iconic possibility; so is Plato. But if we go further back with Nye to Parmenides we discover passionless logic beginning to loose itself from the realities of everyday life but still linked to emotive poetry.

The Parmenidean flight from the world is doomed to fail from the beginning. Parmenides speaks for a world without women, sexual generation, fertility, change, emotions, and flesh. Aristotle and Plato speak for a masculine culture divorced from the life of the household. Ockham and Abelard speak for a patristic Church committed to the absolute authority of a transcendent male god, an exclusively male priesthood, and a theology that equates women with sin and evil. And Frege speaks for the male sanctuary of the German university where men engage in symbolic mortal combat to achieve status, power, and privilege. It seems to me that Nye's feminism gives her the courage to try to read logic, but her reading is not so much strictly feminist, history, or philosophy as it is a sociology of knowledge.

Nye claims that logic flourishes where human community fails: the collapse of the city state, economic crises, disease, hunger, exposure to radically different cultures. If logic demands that we ignore context, circumstances, and personhood when assessing the claims of the logician, reading asks us instead to consider these things carefully.

4 Conclusion

Abstraction (I prefer the concretizing term "generalization") depends on realizing opportunities for producing, publishing, and disseminating ideas in a specialized community of teachers, researchers, and students that extends over a number of continuous generations. If the abstracting ("generalizing") process is carried out under the conditions of social fragmentation (e.g., class divisions and class struggles), and the fragmentation of knowledge (e.g., separating and opposing natural knowledge and political-ethical-moral knowledge), it will generate ideologies of purity. Thus, arguments for purity are not simply consequences of professional autonomy and specialization. They are an imperative of social conflict (including sex, gender, race, and class struggles). The work of scientists consists in great part of rationalizing thought and developing algorithms for the ready application of ideas to the political, economic, and military ends of ruling elites. But the more insulated their work becomes from "external" interests, the more the algorithmic imperative is subordinated to the imperatives of teacher/researcher-student relationships and traditions, the more we find self-consciousness about and the defense of professional autonomy. This leads to the development of ideologies that justify and glorify the separation of hand and brain, or more generally the concept of "...for its own sake". Ideologies of purity are not unrelated to the role of the thinking class as a tool of the established order. Ideologies of purity are a product of the extreme separation of hand and brain. Logic is the extreme consequence of this process.

Notice that there is a tension between the role of scientists as the brains (and to some extent the hands too) of the political, military, and economic interests on the one hand, and their roles as members of a professional community whose autonomy guarantees them certain perquisites, especially a certain degree of apparent independence in the pursuit of their teaching and research goals. The resources at the disposal of scientists for sustaining and defending autonomy, however, are available ultimately at the pleasure of the ruling classes. If the ruling classes decide that autonomy is producing too much independence, and that it is generating too much insularity, criticism, and meager payoffs, it will adjust the distribution of resources accordingly in order to effect a lower autonomy profile in and a more directly profitable relationship with, the community. We have seen just this development unfold in the last half century in the historical centers of academic freedom and tenure in the United States and in the goals of the National Science Foundation.

Theory is related to the speculative thought that accompanies the division of society into social classes. It is in general the special possession of the ruling classes. As a ruling class becomes divorced from basic sustenance labor, its thought patterns become increasingly "speculative". Theses thought patterns correspond to the relatively isolated and insular material conditions of the ruling class lifestyle, or at least the lifestyle of that segment of the ruling class divorced from the day to day activities of gaining, keeping, and expanding control of societal resources. A scholarly subclass within or supported by the ruling class transforms speculation to theory. This sub-class is directly or indirectly dependent on the patronage of the ruling class. This distinction between speculation and theory depends in part on the degree to which the theory specialists can and do maintain some ties to productive labor. The ties may be to engineering or applied science, or to the internal labor of science known as "experimentation". The extreme and prolonged separation of the hand and brain endangers the autonomy and the very existence of a community of specialists, as I will argue below.

The development of a more or less autonomous institutional sphere creates a new context for productive labor. The same sort of analysis that at a more primitive level begins with activities aimed at satisfying basic human needs must now be carried out simultaneously on a new level where one aim is to satisfy organizational and institutional imperatives. The institutionalization of science creates the conditions for generating science out of science. This does not mean that the material foundations of earlier sciences are transcended. Rather, they become transformed and more complex. New organizational levels are built upon lower level sustenance organizations. Occupations, professions, and institutions are higher order material foundations for human productive activity.

When we focus our attention on internal social structure and autonomy, it becomes clear that processes of specialization, routinization, institutionalization, professionalization, and bureaucratization increase the degree of closure in a social activity relative to other social activities. As closure increases, the boundary separating a given activity from other activities becomes thicker and more difficult to communicate across. A boundary can be thickened, for example, by increasing the degree of specialization and uniqueness in linguistic, symbolic, and notational systems. The process of closure is initially promoted by large scale developments (for example, increases in the scale of exchange economies). As specialists emerge, they take an increasingly active and self-conscious role in promoting and protecting closure (that is, "boundary work" takes up more and more of their time). Under such conditions, activities within and across generations (assuming generational continuity) will yield increasingly general ("abstract") productions, or objects. Historically, in the absence of the sort of analysis and awareness reflected in this essay, the process of closure eventually leads to conjectures about cultural creation or social production as matters of pure mental activity. As closure proceeds, such conjectures become increasingly prominent and plausible. This is so in part because most workers are ignorant of, or forget, their history and therefore the material, practical, and social roots of their productive activities and their products; and in part because certain more self-aware representatives of the specialty *deliberately* set out to protect a "sacred" image of their work and to compete against other specialists for scarce resources. Purity has both sacred and demarcation functions.

As a social activity becomes more specialized and more autonomous in relation to other social activities, it becomes increasingly focused on its own products as objects *and* tools of social production and reproduction. Given generational continuity and a parasitical or otherwise umbilical relationship to the "external" society (so that significant resources and energy do not have to be devoted to primary sustenance concerns), an iterative process occurs in which the products of one set of activities or of one generation become the materials (the material foundation) for the next set of activities or of the next generation's productive activity carries this process to its extreme and gives logic dominion over the rules of reason in every arena of life. Logic does not exhibit a logical history. It is a proper subject matter for the sociologist of knowledge. And Andrea Nye demonstrates that there is more than one "strictly" sociological path to that conclusion. Nye, the feminist, the philosopher (in unacknowledged concert with the sociologist of knowledge) argues that the genesis of logic(s) *is* relevant to the truth(s) and falsity(ies) of logic. A critical understanding of the logician as a person in a sociocultural context is relevant when assessing whether his or her logical claims are true or false. The goal of the logician is to establish laws of truth telling that stand apart from people, social and cultural conditions, history, and time and space themselves. I have offered some empirically grounded reasons for why at the end of the day logic cannot stand apart from the contingencies of human life and natural language.

References

Amann, K. and K. Knorr Cetina (1989). "Thinking Through Talk: An Ethnographic Study of a Molecular Biology Laboratory". In: *Knowledge and Society: Studies in the Sociology of Science Past and Present.* Ed. by R. A. Jones, L. Hargens, and A. Pickering. Vol. 8. Greenwich, CT: JAI Press, pp. 3–26.
Aristotle (1984). *The Complete Works (revised Oxford translation). Vol. 1.* Ed. by J. Barnes. Princeton: Princeton University Press.
Bloor, D. (1987). "Review of The Living Foundations of Mathematics by Eric Livingston". In: *Social Studies of Science* 17.2, pp. 337–358.
Boole, G. (2009 (1854)). *An Investigation of the Laws of Thought.* Cambridge, MA: Cambridge University Press.
Cairns, H. (2005). "Introduction". In: *The Collected Dialogues of Plato, Including the Letters (Bollingen Series LXXI).* Ed. by E. Hamilton and H. Cairns. 7th ed. Princeton: Princeton University Press.
Clark, G. H. (1980). "God and Logic". In: *The Trinity Review* nov-dec, pp. 52–56.
Collins, R. (1998). *The Sociology of Philosophies.* Cambridge, MA: Harvard University Press.
Douglas, M. (1986). *How Institutions Think.* Syracuse, NY: Syracuse University Press.
Edge, D. and M. Mulkay (1976). *Astronomy Transformed: Emergence of Radio Astronomy in Britain.* New York: John Wiley & Sons.

Fleck, L. (1979 (1935)). *The Genesis and Development of a Scientific Fact*. Ed. by T. J. Trenn and R. K. Merton. Chicago: University of Chicago Press.

Foucault, M. (1972). *The Archaeology of Knowledge & The Discourse on Language*. New York: Pantheon Books.

Goffman, E. (1974). *Frame Analysis: An Essay on the Organization of Experience*. Boston: Northeastern University Press.

Kleene, S. C. (2009 (1950)). *Introduction to Metamathematics*. New York: Ishi Press.

Lotze, R. H. (1843). *Logik*. Leipzig: Weidmann'sche Buchhandlung.

Nye, A. (1990). *Words of Power: A Feminist Reading of the History of Logic*. New York: Routledge.

Restivo, S. (1992). *Mathematics in Society and History. Sociological Inquiries*. Dordrecht: Kluwer.

Rosental, C. (2008). *Weaving Self Evidence: A Sociology of Logic*. Princeton: Princeton University Press.

Shapin, S. and S. Schaffer (1985). *Leviathan and the Air-Pump: Hobbes, Boyle, and the Experimental Life*. Princeton: Princeton University Press.

Struik, D. J. (1986). "The Sociology of Mathematics Revisited". In: *Science and Society* 50, pp. 280–299.

Weisheipl, J. (1970). "Preface". In: *Commentary on the Posterior Analytics of Aristotle, by Thomas Aquinas*. Ed. by J. Kenny. Magi Books.

CHAPTER 8

Mysterianism Revisited: On The Semiotics of Consciousness

Roger Vergauwen

1 Introduction

In a paper on the use of analogy and metaphor in mathematics, Jean Paul Van Bendegem writes the following: "It may perhaps sound strange if not bizarre to suggest that metaphors and analogies could and should play a role in the practice of mathematics, let alone the claim that they are essential in present-day mathematics. Yet, that will be precisely the claim I will defend in this paper ..." (Van Bendegem 2000, p. 105). After having argued successfully in favor of this, he concludes the paper thus: "If it is indeed the case that analogies and metaphors are essential to mathematics, then it becomes quite reasonable to suppose that such a thing as a *semiotics* of mathematics becomes possible, if not necessary" (Van Bendegem 2000, p. 122). In this paper, we will investigate why such a semiotics isn't only useful in mathematics but also in the philosophy of mind. In a first part we will discuss Colin McGinn's formulation of the mind-body problem and his proposed solution which has sometimes been called *Mysterianism* (Flanagan 1992). According to McGinn, the mind-body problem cannot be solved because of our innate inability to do so, something he calls *cognitive closure*. In this argument, McGinn uses several analogies that are supposed to prove his point. In a second part these analogies will be examined and it will be shown that they are deficient in several ways. Moreover, as we will try to show in the next part using ideas in the philosophy of language of Michael Dummett, it will turn out that McGinn's cognitive closure is more related to the realist semantics of the language used to discuss the mind-body problem than to the ontology of the subject matter. It might, then, make sense to talk about McGinn's mysterianism as analogical to a problem in the syntax and semantics of language —and so of semiotics— rather than merely as an epistemological (or ontological-conceptual for that matter) one. Finally, it will be suggested that McGinn's mysterianism can be reformulated in such a way that it becomes less intractable and less mysterious.

2 McGinn's Answer to the Mind-Body Problem

The problem of consciousness has been generally recognized as the hard problem of the mind-body problem. Dualism, materialism, idealism, have all produced their own answers and solutions, but the problem and the mystery remain. McGinn summarizes the situation as follows:

> How is it possible for conscious states to depend upon brain states? How can technicolour phenomenology arise from soggy grey matter? What makes the bodily organ we call the brain so radically different from other bodily organs, say the kidneys-the body parts without a trace of consciousness? How could the aggregation of millions of insentient neurons generate subjective awareness? We know that brains are the de facto causal basis of consciousness, but we have, it seems, no understanding whatever of how this can be so. It strikes us as miraculous, eerie, even faintly comic. Somehow, we feel, the water of the physical brain is turned into the wine of consciousness, but we draw a total blank on the nature of this conversion. (McGinn 1989, p. 349)

As Chalmers has put in relation to the problem of qualia and consciousness: "If any problem qualifies as *the* problem of consciousness, it is this one. In this central sense of 'consciousness', an organism is conscious if there is something it is like to be that organism, and a mental state is conscious if there is something it is like to be in that state. Sometimes terms such as 'phenomenal consciousness' and 'qualia' are also used here" (Chalmers 1995, p. 201). Levine 1983 has called this the problem of the *Explanatory Gap*.

2.1 Transcendental Naturalism

McGinn's solution to this problem, which has been dubbed *Mysterianism*, because it relies on the idea of a mystery being a question that happens to fall outside of a given creature's cognitive space (McGinn 1993, p. 3), starts from what he calls transcendental naturalism. According to transcendental naturalism (McGinn 1993, pp. 3–8) about a certain question Q with respect to a certain being B, it may be the case that the answer to that question is beyond the (cognitive) capacities of such a being. That is, the answer to a certain question may be epistemically inaccessible to a certain being. This is what he calls cognitive closure. "A type of mind M is cognitively closed with respect to a property P (or theory T) if and only if the concept-forming procedures at M's disposal cannot extend to a grasp of P (or an understanding of T)" (McGinn 1989, p. 350). McGinn maintains that consciousness is a natural phenomenon, arising from the organizations of matter (McGinn 1989, p. 353) , but at the same time advocates epistemic irreducibility, being the view that holds that there is no explanation of consciousness available to us because of the cognitive closure of the human mind. The situation is not unlike the one described Nagel's famous example of bats. A man born blind cannot really understand what it is to have visual experiences and human beings can not by the same token understand what it means for a bat to subjectively experience echolocation. However, even though from an objective point of view we can investigate empirically the physiological details which cause the supposed experiences of bats, it is not open to us to experience *what it would be like to* have these subjective experiences: "For if the facts of experience —facts about what it is like for the experiencing organism— are accessible only from one point of view, then it is a mystery how the true character of experiences could be revealed in the physical operation of that organism. The latter is a domain of objective facts par excellence-the kind that can be observed from many points of view and by individuals with differing perceptual systems" (Nagel 1972, p. 442). According to Nagel the problem is epistemic rather than ontological, because it implies a problem for physicalism: "Physicalism is a position we cannot understand because we do not at present have any conception of how it could be true" (Nagel 1972, p. 446). According to McGinn, however, the problem goes even deeper. For, suppose that, in the case of consciousness, there is a certain property P, which is instantiated in the brain and which is the causal basis of consciousness, and there would also be some theory T, referring to P, that would explain the causal nexus between conscious states and brain states, then of course if we knew that theory we would have a solution to the mind-body problem. However, according to McGinn, it will never be possible to have such a theory and to therefore acquire an understanding of the nature of P. The reason for this is that we are cognitively closed towards an understanding of T: "Rather as traditional theologians found themselves conceding cognitive closure with respect to certain of the properties of God, so we should look seriously at the idea that the mind-body problem brings us bang up against the limits of our capacity to understand the world" (McGinn 1989, p. 354). McGinn's argument therefore, takes the form of a reductio (see also Rowlands 2010, p. 337 and McGinn 1989, pp. 354–359):

(1a) There exists a property P, which is instantiated in the brain, and which is the causal basis of consciousness.

(1b) Identifying this property can be done in basically two different ways, viz. direct investigation through introspection or empirical study of the brain.

(1c) Direct investigation through introspection cannot identify P.

(1d) Emprical study of the brain cannot identify P.

(1e) Therefore, P cannot be identified.

2.2 Conceptual and Cognitive Closure

That direct investigation through introspection cannot be used to identify the required property seems evident. It is of course true that we have immediate access to our own consciousness and its properties, but this does not imply that the required property could be found in this way because by itself this introspection does not reveal the nature of P. Introspection, McGinn argues, reveals our experience to us but it does not reveal to us how conscious states depend upon brain states. An analysis of the concept of consciousness by means of some procedure of conceptual analysis will not do either because by such an analysis we will not be able to see how consciousness depends upon the brain any more than we could solve the problem of how life depends upon physical properties by reflecting on the concept of *life*. Therefore, in as far as introspection could be used as a tool to construct concepts with respect to the nature of P, it is conceptually closed with respect to P.

But, according to McGinn, an empirical investigation of the brain cannot be of help either if we want to understand the nature of P. He wants to argue that P is also perceptually closed to us: "The argument for perceptual closure starts from the thought that nothing we can imagine perceiving in the brain would ever convince us that we have located the intelligible nexus we seek. No matter what recondite property we could see to be instantiated in the brain we would always be baffled about how it could give rise to consciousness ... Inference to the best explanation of purely physical data will never take us outside the realm of the physical, forcing us to introduce concepts of consciousness" (McGinn 1989, p. 357–358). Our senses are geared towards a representation of the spatial world and do therefore represent things as essentially existing in space and therefore with spatial properties (McGinn 1989, p. 363 n21). Since consciousness is not spatially defined, it falls outside the realm of our observational capacities and it cannot be grasped by our spatially-oriented senses. It is, then, because the empirical property that is supposed to provide the causal nexus between the brain and the mind is *noumenal* with respect to our senses that we will never be able to provide an answer to the mind-body problem. Furthermore, when we introduce theoretical concepts (unobservables) on the basis of observation, such an inquiry always obeys a principle of homogeneity which is based on analogical extensions of what we observe. For example, when when we first want to introduce the concept of an atom or molecule we will introduce this by first considering perceptual representations of larger objects and then conceive of smaller objects of the same kind (McGinn 1989, p. 358). According to McGinn, this analogical procedure does not work in the case of consciousness since analogical extensions of properties we find in the brain that are all spatial will not suffice to solve the emergence of consciousness from the physical brain: "Since analogical extensions of the entities we observe in the brain are precisely as hopeless as the original entities were as solutions to the mind-body problem. We would need a method that left the base of observational properties behind in a much more radical way" (McGinn 1989, p. 359). It follows, then, that the property or the set of properties that would be the physical basis of consciousness are cognitively closed to us. An important reason for McGinn to conclude to our cognitive closedness to certain properties, has to with the limits of using an analogical extension (an analogical procedure) in our inquiry into the nature of empirical properties such as in the study of consciousness. Furthermore, it is just because of this that, according to McGinn, consciousness looks like something epiphenomenal:

> This is, indeed, why it seems that consciousness is theoretically epiphenomenal in the task of accounting for physical events. No concept needed to explain the workings of the physical world will suffice to explain how the physical world

produces consciousness. So if P is perceptually noumenal, then it will be noumenal with respect to perception-based explanatory inferences. Accordingly, I do not think that P could be arrived at by empirical studies of the brain alone. (McGinn 1989, p. 359)

When it comes to the question of the strength of his cognitive closure thesis with respect to the mind-body problem, McGinn makes a distinction between absolute and relative claims of cognitive closure. It is held that a problem is absolutely cognitively closed if no possible mind could dissolve it; a problem is relatively closed if minds of some sorts can in principle solve it while minds of other sorts cannot. With respect to the mind-body problem McGinn seems to opt for the first possibility. Since all (empirical) concept formation is tied to perception and introspection it will be highly improbable, for reasons explained above, that we would ever be able to discover the causal nexus between the mind and the brain. Perception and introspection are just not the right categories by means of which that causal nexus could be elucidated. However, McGinn does not want to exclude the second possibility either. And here he makes yet another analogy, this time to concept formation in mathematics. Indeed, "our mathematical concepts (say) do not seem tied either to perception or to introspection, so there does seem to be a mode of concept formation that operates without the constraints I identified earlier" (McGinn 1989, p. 361). It is, then, a kind of a priori reasoning that might come to the rescue here. A mind that could conceive of the mind-brain relation "in totally a priori terms ... Such a mind would have to be able to think of the brain and consciousness in ways that utterly prescind from the perceptual and the introspective-in somewhat the way we now (it seems) think about numbers ... Perhaps this is how we should think of God's mind, and God's understanding of the mind-body relation" (McGinn 1989, p. 361). This last line is probably meant ironically, since McGinn does not think that there is a scientific solution to the mind-body problem. In his view, consciousness has an explanation "in a certain science, but this science is inaccessible to us as a matter of principle" (McGinn 1989, p. 362). Whatever else this science may be like, it is clear that at least one principle will probably have it be upheld which is the principle of causality already well defined by Hume: "... All arguments concerning existence are founded on the relation of cause and effect; that our knowledge of that relation is entirely from experience; and all of experimental conclusions proceed upon the supposition that the future will be conformable to the past" (Hume 1777, pp. 36–37). Even if McGinn has a very different science in mind, causality in some sense will have to be present there and the mind-brain relation can, therefore, not be totally a priori. We will come back to this later. However, McGinn has even a third analogy up his sleeve, more specifically when he relates his transcendental naturalism to a more general theory of human cognitive capacity.

2.3 CALM, Compositionality, Recursivity and the Martian Argument

The issue of there being a science which is in principle inaccessible to us and thus beyond our conceptual powers has, according to McGinn, to do with inherent limitations on our knowledge-acquiring faculties. And here McGinn relies on a Chomskyan background (McGinn 1993, p. 25 n.11). He considers his theory of human cognitive capacity and reason as an analogue of a theory of the universal structure of human language such as proposed by Chomsky (McGinn 1993, p. 18).

2.3.1 THE CALM CONJECTURE

Analogous to what Chomsky says about the language faculty which has been formed in a biological and evolutionary process and is therefore contingent given that the grammar of human languages determines the scope and limits of the human language faculty, McGinn wants to maintain that the same holds for the faculty of reason: "What transcendental naturalism ideally requires, then, is something to play the role of grammar in delimiting what is accessible to reason, where this something fixes boundaries across which philosophical thought cannot travel" (McGinn 1993, p. 18). In his view there is no reason for thinking that we are able to understand everything that there is to be understood about the natural world and that there are limits to what is accessible to reason. McGinn argues that it is the non-spatial character of consciousness which lies at the heart of the problem.

It is here that he develops what he calls the CALM Conjecture. The acronym stands for 'Combinatorial Atomism with Lawlike Mappings'. The basic idea is that some kind of reality can only be understood if we are able to break it down into simpler and simpler elements until some basic level is reached:

> This combinatorial mode of thought, which yields a certain kind of novelty in the domain at issue, and proceeds in bottom-up style, may represent contemporaneous relations between the structures dealt with, as well as dynamic relations over time. The essence of it is to yield understanding

of the domain, especially in generative aspects, by means of transparent relations of composition between elements ... a CALM theory tells us what the nature of these relations is— it specifies the manner in which the domain is structured. To grasp the theory is thus to understand the domain. (McGinn 1993, p. 19)

McGinn sees applications of this principle in linguistics, where we construct complex sentences out off primitive elements such as words and where the properties of these sentences are determined by the properties of the elements they consist of both syntactically and semantically, and where the linguistic structure can be generated starting from a finite base of elements, in mathematics (number theory) and set theory where functional relations are present everywhere. It follows from this conjecture that the entities that conform to this conjecture can be understood, whilst others that do not conform, cannot. Consciousness, being non-spatial, does not conform to this conjecture and therefore cannot be understood:

> Conscious states are not CALM-construable products of brain components. Here the mappings, which must exist in some form, are inscrutable in CALM terms. We can readily conceive of higher-level brain functions in terms of simpler composing constituents; but once we think in terms of consciousness this mode of explanation lapse. (McGinn 1993, p. 37).
>
> In short, the CALM structure is to philosophical problems what human grammar is to nonhuman languages- an unavoidable but unsuitable mode of cognition. (McGinn 1993, p. 20)

2.3.2 COMPOSITIONALITY AND RECURSIVITY AS SEMANTIC PRINCIPLES

The reference to Chomsky and linguistics in McGinn also shows something else. What McGinn's CALM conjecture really comes down to is nothing but the (Fregean) *Principle of Compositionality*. The principle of compositionality in one of its formulations (Janssen 1997, p. 419) with respect to the semantics of languages holds that:

(2) The meaning of a compound expression is a function of the meanings of its parts.

It implies that the meaning of a complex expression is constructed from the meanings of its composing parts by means of certain (semantical) rules. Such a semantics is, then, also called a compositional semantics. A compositional semantics uniquely determines the semantic value of complex expressions based on the semantic values of their immediate parts and on the ways these parts are combined. Though they are methodologically different, the notion of compositional semantics is often identified with that of a recursive semantics and a classical example of such a semantics is the semantics of classical propositional logic (Sher 2001, p. 202). A compositional or recursive semantics allows us to construct an infinite number of sentences from a finite base and by finite means in such a way that it is always clear and uniquely determined how the meaning of the more complex depends upon the less complex. It puts a constraint on the semantics of a language in its relation to the syntax of that language. It is, therefore, a methodological notion. However, to Chomsky —as for McGinn— and others, this methodological notion has much deeper consequences since it is a unique feature of human language and human communication. It is a property which has been evolved evolutionary and is unique to the human species. In a somewhat provocative paper Chomsky a.o. (Hauser, Chomsky, and Fitch 2002) make clear why this is so. They make a distinction between a faculty of language in the broad sense (FLB) and a faculty of language in the narrow sense (FLN) (Hauser, Chomsky, and Fitch 2002, pp. 1570–1571). The first of these faculties contains an internal computational system in combination with at least two other organism-internal systems which they call the "sensory-motor" and the "conceptual-intentional". The second faculty is the abstract linguistic computational system by itself which is contained in the first faculty but which is independent of the other systems with which it nevertheless interacts. As such, then, the second faculty is a component of the first one and the mechanisms that underlie the one are a subset of the ones that underlie the other. It is in this process that they see a central place for recursion:

> We assume , putting aside the precise mechanisms, that a key component of FLN (faculty of language in the narrow sense) is a computational system (narrow syntax) that generates internal representations and maps them into the sensory-motor interface by the phonological system, and into the conceptual-intentional interface by the (formal) semantic system ... All approaches agree that a core property of FLN is recursion, attributed to narrow syntax in the conception just outlined. (Hauser, Chomsky, and Fitch 2002, p. 1571)

Though most researchers, but not all (Coolidge, Overmann, and Wynn 2011), agree that this is indeed the case, especially when it comes to compositionality in the semantics of natural languages, the principle is not without its problems and has been hotly debated (Janssen 1997, pp. 437–447; Sher 2001). Hintikka's approach to the semantics of natural languages provides an example of this (Hintikka 1975, 1984; Kirkham 1997, p. 244): From propositional logic we know that the operators that are included there are truth-functional. The meaning (truth value) of e.g. a complex proposition which has been constructed by means of the implication-operator is clearly a function of the meaning of the composing propositions. But consider the following sentence:

(3a) Any corporal can become a general.

In this sentence the quantifier 'any' is to be understood as a universal quantifier, and the sentence will therefore be true if and only if every corporal can become a general. However, consider sentence (3b):

(3b) If any corporal can become a general, then I'll eat my hat.

In the interpretation of (3b), where 'any' follows if, it is to be interpreted as an existential quantifier in the sense that the sentence is true if and only if even a single corporal can become a general. Hence, it follows that the truth conditions of a compound sentence are not always a function of the truth conditions of its components. Hintikka has, as a consequence of these problems, in later work (Hintikka 1996; Hintikka and Sandu 1997) subsequently developed a game-theoretic semantics which contains a non-compositional, game-theoretic definition of truth for specific kinds of languages (Sher 2001, p. 212).

An even more dramatic example where compositionality is threatened is raised by belief-sentences and the substitution of synonyms in belief-contexts (Janssen 1997, pp. 444–445). To illustrate this, let us compare the following sentences:

(4) John believes that Alfred is a child Doctor.

(5) John believes that Alfred is a pediatrician.

Sentence (4) may well be true while sentence (5) is false, despite the fact that 'child Doctor' and 'pediatrician' are synonymous. From the fact that John believes what is expressed in sentence (4) it does not follow that he also believes what is expressed in (5). However, if compositionality holds and the embedded sentences in (4) and (5) are synonymous, it should be the case that if John believes (4) then he also believes (5).

On the other hand it is also obvious that compositionality or a compositional approach has certain advantages, such as presenting an account of the learnabilty, novelty, productivity, systematicity and intersubjectivity of languages (Pagin and Westertahl 2010, pp. 265–270). The debate on the necessity and possibility of a compositional semantics for natural languages is still going on and it seems safe to say that it will be with us for a long time to come. It is not the aim of this paper to make a final judgment on these matters, but it seems that Chomsky's realism about the nature of linguistic mechanisms and McGinn's analogous realist interpretation in the CALM conjecture as ontological are responsible for at least part of the trouble with mysterianism. In order to begin to see why this is so, it might be useful to consider an argument —the Martian Argument— against Chomsky's theory which explicitly aims at this realism.

2.3.3 CALM ON MARS? LINGUISTIC COMPETENCE AND THE MARTIAN ARGUMENT

Some have argued that Chomsky's approach, who considers linguistics to be part of cognitive psychology, is misguided. Devitt and Sterelny have presented an argument —dubbed the Martian argument by Stephen Laurence (Laurence 2003)— intended to show why this is so (Devitt and Sterelny 1989). Linguistics, according to Chomsky, deals with human knowledge and understanding of language, linguistic competence, and is therefore concerned with questions about the cognitive skills of human beings and thus in the end ultimately about the specific human neural set up. The psychological mechanisms for language acquisition are instantiated into mind/brains of the human language users. Since grammars are about such instantiations, the instantiation of English grammar in humans is indeed English according to this conception. This we may call the Competence Thesis, the thesis that grammars are about the human language competence. The Martian argument now runs as follows. Let us assume that Martians, whose psycholinguistic processes ex hypothesi differ from human ones nevertheless manage to produce a set of sentences that are extensionally equivalent to the set of sentences in human English. This implies that the sentences that are grammatical in Martian English are also grammatical in regular English. However, since by hypothesis Martians had a different neural organization they really have a different linguistic competence, the question arises whether they should count as speaking English?. On an account of what it is to speak English the Martian speakers should indeed count as speaking English. On the level of linguistic symbols everything they say is indistinguishable from us and they also seem to be able to communicate via a shared language. They are, therefore, on the face of it, competent in English. We can, then, nevertheless study the shared language that we both use on the linguistic level without having to appeal to differences in competence:

> According to the transformationalists, English competence consists in internalizing a grammar. They go further: all English speakers have internalized near enough the one grammar; competence has a uniform structure across the linguistic community. Even if this is so, it is not necessarily so. Many other grammars could agree on the meaning-relevant structures they assign to the sentences of English. Suppose that Martians became competent in English by internalizing one of these grammars. The theory of Martian competence would have to be different from the theory of ours. Yet the theory of symbols would be the same, for it would still be English that they spoke. Returning to earth, it would not matter a jot to the theory of symbols if competence among actual English speakers was entirely idiosyncratic. In sum, linguistic competence, together with various other aspects of the speaker's knowledge, produces linguistic behavior. That behavior, together with the external environment, produces linguistic symbols. A theory of symbols is not a theory of competence. (Devitt and Sterelny 1989, p. 514)

This argument has caused a lot of discussion (e.g. Borge 2006, Laurence 2003) which we do not want to go into here, but it seems that it has a certain intuitive plausibility. It might indeed be the case that the Martian English speakers have a different kind of linguistic competence which nevertheless allows them to produce the same set of sentences that humans can with their own competence. Such a situation would not prevent us from studying language and its properties, without an explicit knowledge of what exactly the linguistic competence consists in or which mechanisms are involved. It may be the case that Martian English has a Hintikka-style game-theoretic semantics, which is not compositional, while Earth-English may have a compositional semantics. Or it may be the other way around, or even none of the aforementioned languages may have a compositional semantics but only a Hintikka-style game-theoretic semantics. Several kinds of semantics may be empirically equivalent with respect to both Martian and Earth-English, and we would still be able to study the properties of both languages as symbol-systems. We might also, further, speculate that Martians-whose language is by hypothesis not compositional, are beings for which McGinn's CALM conjecture does not hold but that would e.g. not mean that they would not be conscious beings and indeed it would be fully possible to study their consciousness on the basis of their behaviour. Dennett comes close to such an approach (Dennett 1991). According to Searle, Dennett's work in the philosophy of mind is firmly in the tradition of behaviorism — the idea that behavior and dispositions to behavior are constitutive of mental states— and verificationism, according to which the only things which exist are those that can be verified by scientific means (Searle 1997, pp. 97–101). The 'stream of consciousness' that is experienced is characterized by a certain ontological subjectivity that goes with phenomenal states. They are such states as "seeing red", "feeling pain", and even "being conscious". Such phenomenal states are called "raw feels" or "qualia". They express the view of "what it is like to be". Dennett, now, thinks that there are no such things as qualia, or subjective first-person phenomena. He claims that it merely *seems to us* that there such things as qualia, but this is a mistake. The only things there really are, are stimulus inputs and dispositions to behavior. He thinks that all data concerning consciousness are explainable from a third-person (objective) point of view and makes hay of theories that postulate uninvestigatable entities:

> Why should it matter if they [zombies] were conscious— especially if consciousness were a property, as some think, that forever eludes investigation? ... Consciousness, you say, is what matters, but then you cling to doctrines about consciousness that systematically prevent us from getting any purchase on why it matters. Postulating special inner qualities that are not only private and intrinsically valuable, but also uncomfirmable and uninvestigatable is just obscurantism. (Dennett 1991, p. 450)

Dennett is here not explicitly criticizing McGinn, nor does it seem to the present author that Dennett's alternative theory is necessarily a better one, but in view of what has been said before it would seem that McGinn's Chomskyan approach and the analogy to the principle of compositionality in his appeal to the CALM conjecture are highly questionable, especially since the principle is interpreted realistically and ontologically and not merely methodologically (even though it is called a conjecture). In this respect, it should be noted that McGinn himself is not completely convinced of the principle either, esp. when applied to the sciences, when he writes: "Quantum physics is theoretically problematic, at least in certain respects, just because it fails of CALM integration" (McGinn 1993, p. 25 n12). Indeed, M. Dummett has even cast doubt on the very possibility of a systematic description of natural language by a theory of meaning, compositional or otherwise: "There can be no guarantee that a complex of linguistic practices which has grown up by a piecemeal historical evolution in response to needs felt in practical communication will conform to any systematic theory" (Dummett 1976, p. 104). As we will see in the next part, an analogous problem shows up when McGinn appeals to the use of analogical procedures in concept-formation.

3 McGinn Dummettized : Realist and Anti-Realist Semantics

As was mentioned before, McGinn's mysterianism turns around the question of understanding how something essentially non-spatial (consciousness) could arise from something essentially spatial. However, according to McGinn, the explanatory property that is required to see this is noumenal for us. But there is such a property nevertheless. "We find it taxing to conceive of the existence of a real property, under our noses as it were, which we are built not to grasp-a property that is responsible for phenomena that we observe in the most direct way possible. This kind of realism, which brings cognitive closure so close to home, is apt to seem both an affront to our intellects and impossible to get our minds around" (McGinn 1989, p. 365). It is this kind of realism which lies at the heart of McGinn's approach to a solution of the mind-body problem, and it is a consequence of the nature of his transcendental naturalism (TN):

> Plainly TN accepts a strong form of realism; in particular, it accepts realism about *the nature* of the things that cognitive beings think and talk about. While we may be able to refer to certain things, there is no guarantee that we shall be able to develop adequate theories of these things. Put differently, the correct theory of what is and is referred to, conceived as a set of propositions detailing the nature of those referents, may not belong to the space of theories accessible to the beings under consideration—including human thinkers. So, for TN, there may exist facts about the world that are inaccessible to thinking creatures as ourselves. Reality is under no epistemic constraint. (McGinn 1993, p. 5)

It follows from this that McGinn is committed to a language the semantics of which is to be interpreted realistically and with sentences possibly having truth-conditions that are (strongly) recognition-transcendent. In the case of the propositions expressing the causal nexus between the brain and the mind, the propositions expressing the spatial causal basis of the non-spatial (consciousness), these propositions would even be recognition-transcendent in principle in McGinn's view, since, among other things, the concepts used would be such that they cannot be formed by analogical extensions of the concepts we are familiar with. However the very idea of such a realist semantics has met with strong criticism, especially from Michael Dummett.

3.1 M. Dummett on Realism in Semantics[1]

Michael Dummett has argued strongly against any realist approach to semantics, not exclusively on logical grounds, like H. Putnam, but in combination with an epistemological criticism of language understanding. For Dummett, realism is a *semantic thesis* as to what makes the sentences of a language *true*:

> So construed, realism is a semantic thesis, a thesis about what, in general, renders a statement in a given class true, when it is true. The very minimum that realism can be held to involve is that statements in the given class relate to some reality that exists independently of our knowledge of it, in such a way that that reality renders each statement in the class determinately true or false, again independently of whether we know, or are even able to discover, its truth-value. This realism involves acceptance, for statements of the given class, of the principle of bivalence, the principle that every statement is determinately either true or false. (Dummett 1982, p. 55)

This view fits well with a *truth-conditional semantics*, which supposes that sentences must have particular truth-conditions by virtue of which they can be *true*. According to Dummett, a truth-conditional semantics as a theory of meaning does not, however, need to have a realist conception of truth, and as a logical theory it does not need necessarily to speak about an independently existing reality which 'makes sentences true'.

Meaning, according to Dummett, even more than being a matter of truth-conditions, is also a matter of *understanding* language:

---

[1] For the materials in this subsection: Vergauwen 1992, pp. 143–151.

> I believe the correct observation that philosophical questions about meaning are best interpreted as questions of *understanding*: a dictum about what the meaning of an expression consists in must be construed as a thesis about what it is to *know* its meaning. So construed, the thesis becomes: to know the meaning of a sentence is to *know* the conditions for it to be true. (Dummett 1976, p. 69)

Now, according to Dummett, this epistemological conception of *knowledge of truth-conditions* is irreconcilable with a realist view on the concept of truth (Grayling 1982, p. 234). A theory of meaning must be capable of explaining what the speakers of a particular language L know, if they know what the sentences of L mean. Therefore, this theory must provide a theoretical representation of the explicit knowledge of the language users and how they apply that knowledge in the use of the sentences of their language. Therefore the knowledge of truth-conditions is of great importance in the *concrete use of the language*. The connection between knowledge of truth-conditions and the actual language use must be clarified because there are at least two conditions that any acceptable theory of meaning has to meet. The first is that in one way or another it must be possible to distinguish what counts as a manifestation of the knowledge which a language user has of the sentences of this language. The second condition is that a theory of meaning- since language is a public means of communication and it is therefore not sufficient to be able to say what individual language users know when they know their language- must be capable of explaining how the *process of learning* a language works. If the process of learning a language is a public affair, then it certainly has to do with the learning or teaching of truth-conditions, and it is in this connection that Dummett thinks it necessary to reject a realist conception of truth. If knowledge of truth-conditions consists in the possibility of knowing when a given sentence is true or false and thus of *recognizing* whether or not *particular* truth-conditions occur there is no problem for the connection between language use and knowledge, because such a recognition has to do with a *practical capability* of recognizing sentences in their use as being *true* via a particular learning process. If truth-conditions, however, are construed as possibly *experience-transcending* properties of sentences, as the realist maintains, how then could an explanation be given for the *knowledge* of such truth-conditions?

For Dummett a central question is the following: "What is it that a speaker knows when he knows a language, and what in particular does he thereby know about any given sentence of the language?" (Dummett

1976, p. 69). No matter which theory it is that wants to give an answer to this, it must contain a theoretical representation of a practical ability, namely the ability to speak a language. This theoretical representation must consist of a set of propositions (theory) and it will be the explicit statement of the implicit linguistic knowledge. But if this theory wants to make the *implicit knowledge* explicit, then it must also be able to specify what the manifestation of this knowledge precisely involves.

> Without this, not only are we left in the dark about the content of ascribing such knowledge to a speaker, but the theory of meaning is left unconnected with the practical ability of which it was supposed to be a theoretical representation. It is not enough that a knowledge of the theory of meaning be said to issue in a practical ability to speak the language: for the whole point of constructing the theory was to give an analysis of this complex ability into its interrelated components. (Dummett 1976, p. 71)

The theory which Dummett has in mind and which must explain linguistic behavior consists essentially of three subtheories, each of which has a specific function. The most important part of such a theory of meaning is a *theory of reference*, which is a truth theory and contains an inductive characterization of the truth values for the sentences of a language: "It is called a theory of reference because, while some of its theorems will state the truth-conditions of sentences, its axioms, which govern individual words, will assign references to those words" (Grayling 1982, p. 273). This theory is supplemented by what Dummett calls a '*theory of sense*', which explains what a language user knows when he knows the reference theory: "A theory of sense will lay down in what a speaker's knowledge of any part of the theory of reference is to be taken to consist, by correlating specific practical abilities of the speaker to certain propositions of the theory" (Dummett 1976, p. 74). Finally, the theory of meaning contains yet another subdivision, the '*theory of force*'. This part of the theory is needed in order to be able to make a classification of the different types of sentences which are used in language: "The theory of force will give an account of the various types of conventional significance which the utterance of sentences may have, that is, the various kinds of linguistic acts which may be effected by such an utterance, such as making an assertion, giving a command, making a request, etc." (Dummett 1976, p. 74). From this three-fold division, it appears that the general picture is one of a theory which identifies knowledge of the truth-conditions for sentences with the understanding of the sentence meaning. The question remains, however, as to whether the *concept of truth* contained in 'truth-conditions' is the right choice as a central concept in the theory of meaning. That truth is a basic concept in the theory of meaning has some measure of plausibility, which in its obviousness, however, is weakened by the question as to which analysis must be given to the truth-concept. In the framework of Dummett's theory, this includes the question as to where the concept of truth itself comes up in the *process of learning* a language. It is clear that this concept cannot be introduced in a stipulative way, because this presupposes that we already know a significant fragment of a language, namely that in which the truth definition would be stated: "If we want to maintain that what we learn as we learn the language, is, primarily, what it is for each of the sentences that we understand to be true, then we must be able, for any given sentence, to give an account of what it is to know this which does not depend upon a presumed prior understanding of the sentence; otherwise our theory of meaning is circular and explains nothing" (Dummett 1976, p. 78). The above need not mean that the meaning of a sentence cannot at all be given in terms of truth-conditions; it is just that this is in principle not possible for the *whole* of the sentences of a language. The real problem in giving a reasonable explanation what precisely the knowledge of the language user consists in in understanding the truth-conditions of sentences is, according to Dummett, not so much that it would not be possible to decide what the *recognition* of a truth-condition consists in, because *if* the truth-condition for a sentence can be given and can be *recognized* , then it is indeed always possible to indicate in principle what it means to (re)cognize these truth-conditions: "That knowledge will consist in the speaker's capacity, perhaps in response to suitable prompting, to evince recognition of the truth of the sentence when and only when the relevant condition is fulfilled" (Dummett 1976, pp. 80–81).

Such an explanation, however, is only valid *for those sentences of which the truth conditions can be effectively recognized,* where the truth-value of the sentence is thus in principle *decidable*: "That is, for which a speaker has some effective procedure which will, in a finite time, put him into a position in which he can recognize whether the condition for the truth of the sentence is satisfied" (Dummett 1976, p. 81). For such sentences it can be said that the knowledge of the truth-conditions consists in the fact that the language user has a particular decision procedure which, under favorable conditions, he can carry out such that after the application of that procedure he can be regarded as understanding a particular truth-condition. The sentences of a language for which this is possible are called the *decidable part* of a language by

Dummett and, as far as the decidable part is concerned, there is thus no distinction between the *truth-conditions* and the *assertibility-conditions* of these sentences. The assertibility conditions are the procedures which a language user can apply in order to determine whether or not a particular truth-condition occurs. According to Dummett, however, besides these decidable sentences, language also contains a lot of *'not effectively decidable' sentences*, thus for which no effective procedure exists for determining, in a finite time, whether or not their truth-conditions are satisfied or can be recognized as true or false. This set contains sentences, such as, for example, references to inaccessible regions of space-time, as well as sentences which require quantification over infinite domains, for example infinite domains of moments of time, the truth-value of which could only be determined by an infinite set of facts. For example:

(6) In this place a city will never be built

If, as the realist says, there is something by means of which sentence (6) can be true, then it can be nothing else than the 'infinite' set of facts that no city will be built on the place intended in 2013, nor in 2014, nor in 2015, ... and so on ad infinitum. Apparently it is not possible in this case to say with certainty that on this place a city will stand in the year n, for a specific n, or to find a general proof that in this place a city will never be built (see also Haack 1974, p. 105). Further, this not effectively decidable fragment of natural language will also contain sentence in which experiences are ascribed to persons and sentences with existential quantification such as (7).

(7) Somewhere in the universe there is a little green Hobbit.

As with (6), knowing the truth-condition of (7) would require checking an infinite number of places in the universe, which is not effectively possible. For sentences which in this way are not effectively decidable, it may indeed be possible to find out if their truth-condition is *in fact* fulfilled or not, but the knowledge of this cannot in general be equated with the knowledge of their truth-conditions: For such a sentence, we cannot *equate* a capacity to recognize the satisfaction or non-satisfaction of the condition for the sentence to be true with a knowledge of what that condition is (Grayling 1982, p. 239). Dummett states:

> [We cannot make such an equation] because, by hypothesis, either the condition is one which may obtain in some cases in which we are incapable of recognizing that fact, or it is one which may fail to obtain in some cases in which we are

incapable of recognizing that fact, or both. Hence, a knowledge of what it is for that condition to hold or not to hold is, while it may demand an ability to recognize one or another state of affairs whenever we are in a position to do so, cannot be exhaustively explained in terms of that ability. In fact, whenever the condition for the truth of a sentence is one that we have no way of bringing ourselves to recognize as obtaining whenever it obtains, it seems plain that there is no content to an ascription of implicit knowledge of what that condition is, since there is no practical ability by means of which such knowledge may be manifested. (Dummett 1976, pp. 81–82)

The problem that appears here is connected with the *theory of sense*, the aim of which is to determine the knowledge which lies bound up in the *theory of reference*. What a language user learns when he learns to speak his language is a *particular practice*, part of which includes learning to recognize a sentence as *true* or *false*. In other words, he learns to do and to say different things as an expression of such a recognition. What a language user knows, what enables him to use his language, must be manifest in the practice itself of speaking, but "knowing the condition which has to obtain for a sentence to be true, is not anything which he *does*, nor something of which anything that he does is the direct manifestation" (Dummett 1976, pp. 82–83). Thus, although in some cases it is possible to ascribe knowledge of truth-conditions to a language user, this is not always possible, and especially not in the case of the not effectively decidable sentences. Dummett's conclusion is thus that knowledge of truth-conditions can offer no explanation for knowledge of the meaning in general, certainly not if it must be taken into account that, as is apparent from the foregoing, truth-conditions can be *recognition transcendent*, and therefore not effectively decidable.

Dummett's *antirealist alternative* then consists, just as with e.g. Hilary Putnam (Putnam 1980), in presenting a semantics in which truth-conditions are not central and not every sentence is true or false (bivalence). He finds a prototype of this in intuitionism, the fundamental idea of which is that our understanding of a mathematical sentence is based on the possibility of recognizing whether a particular construction constitutes a *proof* of the sentence or formula. In this way we can be certain that the understanding of an expression has to do exclusively with the use, because *recognizing* of a proof has to do directly with the use of symbols. Just as little does this view require that each sentence be effectively decidable, because 'understanding' does not consist in the finding of a proof, but in the recognition of a construction *as a proof*, if a proof exists and is given: "Our understanding of a statement consists in a capacity, not necessarily to find a proof, but only to recognize one when found" (Dummett 1976, p. 110). According to Dummett such a theory, which repeats Wittgenstein's dictum 'Meaning is Use', is also suitable in the semantics of natural language, in which the *verification principle* stands central: "Such a theory generalizes readily to the non-mathematical case. Proof is the sole means which exists in mathematics for establishing a statement as true: the required general notion is, therefore, that of *verification*. On this account, an understanding of a statement consists in a capacity to recognize whatever is counted as verifying it, i.e. as establishing it as true" (Dummett 1976, pp. 110–111). From this point of view a sentence can indeed have a meaning, but no truth-conditions. Dummett admits that it is not easy to account for the notion of truth in terms of a verificationist theory of meaning, but he maintains that it must be explained in terms of our capacity to recognize statements as true and not in terms of recognition-transcendent truth-conditions. This is also exactly what Putnam had in mind when he introduced the concept of *internal realism*:

> Objects in constructive mathematics are given through descriptions. Those descriptions do not have to be mysteriously attached to those objects by some non-natural process (or by metaphysical blue). Rather, the possibility of proving that a certain construction (the 'sense', so to speak, of the description of the model) has certain constructive properties is what is asserted and all that is asserted by saying the model 'exists'. In short, *reference is given through sense, and sense is given through the verification-procedures and not through truth-conditions.* The 'gap' between our theory and the 'objects' simply disappears-or, rather, it never appears in the first place. (Putnam 1980, p. 479).

Dummett's most important criticism, through which he came to reject realism, is, then, the existence of *not effectively decidable sentences* in a language. Only in the case of not effectively sentences does the realist seem forced to resort to a concept of truth which is recognition transcendent: "Put differently, the anti-realist will object to the realist only with respect to sentences for which the latter claims that truth conditions and assertibililty conditions diverge in some way" (McGinn 1982, p. 119).

3.2 Transcendental Naturalism and Effective Decidability

McGinn has criticized Dummett's global anti-realist semantics as being inconsistent and reductionist (McGinn 1979). However this may be, from the above it becomes clear that McGinn's position in relation to the mind-body problem can be formulated using the concept of not effectively decidable sentences. Furthermore, it can be shown that it is precisely McGinn's view of a realist interpretation in the semantics of his own Transcendental Naturalism that is a further basis for his mysterianism, since it crucially makes use of the concept of recognition-transcendent truth-conditions. This position, then, may be summarized as follows (McGinn 1989, pp. 352–362):

(8a) There exists a property, P, instantiated by the brain, in virtue of which the brain is the basis of consciousness. In particular, P provides the causal basis of consciousness.

(8b) P is the spatial causal basis of the non-spatial (consciousness).

(8c) We are (probably absolutely) cognitively closed to P or to a theory which contains P.

(8d) P has a full and non-mysterious explanation in a certain science that is inaccessible to us as a matter of principle.

(8d) indicates that the problem involved is also an epistemological one, in the sense that we are not required to abandon our materialist intuitions to acknowledge a mysterious non-material mind. So, the mystery is not metaphysical. However, at the same time (8a) is a proposition which, because of (8c-d), has recognition-transcendent truth-conditions (for human beings) and can, then, be said to be a not effectively decidable sentence in the Dummettian sense. It is not effectively decidable from our perspective and our science. In McGinn's view, the proposition (8a) is nevertheless also true and even effectively decided (8d). The reason for this is that McGinn's transcendental naturalism implies that if a sentence is true, there must be something in virtue of which it is true- a principle which underlies the correspondence theory of truth- and a realist interpretation of the semantics of language. As Dummett puts it: "We may, in fact, characterize realism concerning a given class of statements as the assumption that each statement of that class is determinately either true or false" (Dummett 1976, p. 93). For Dummett, non-effectively decidable sentences are not stably true or false and this is also why he rejects the principle of bivalence for propositions.

3.3 Science, Effective Decidability and Truth

But why does McGinn think that (8a) is true? The answer is surprisingly simple, albeit not unproblematic:

> The philosophical problem about consciousness and the brain arises from a sense that we are compelled to accept that nature contains miracles-as if the merely metallic lamp of the brain could really spirit into existence the Djin of consciousness. But we do not need to accept this: we can rest secure in the knowledge that some (unknowable) property of the brain makes everything fall into place. What creates the philosophical puzzlement is the assumption that the problem must somehow be scientific but that any science *we* can come up with will represent things as utterly miraculous. (McGinn 1989, p. 362)

So, in fact, it is his (belief in) *naturalism* which forces McGinn to accept the truth of (8a). (8a) is effectively decidable (and even effectively decided), according to McGinn, but only in a science that is inaccessible to us, since we are cognitively closed to it. As was mentioned before, the reason for this is, amongst other things, that the concepts we use cannot analogically be extended to encompass the required property:

> Brain states and conscious states fall under senses such that (i) under those senses the link between them is intelligible and (ii) those senses are not potential constituents of human thought ... They correspond to the kind of conceptual shift that would render the psychophysical relation transparent, *if* it could be achieved. They occur in the propositions that constitute the (ideal) scientific theory of mind and body ... TN (Transcendental Naturalism) thus diagnoses the character of the philosophical problem for us as consisting in the cognitive inaccessibility of the right senses, the ones that convert the problem into regular science. (McGinn 1993, p. 40).

In the case of consciousness it is McGinn's conviction that, since our conceptual and scientific frameworks are thoroughly permeated by concepts that are essentially spatial, these concepts are unsuitable to study the nature of consciousness, which is nonspatial. Such a study would require "leaving behind" the spatial framework of our thinking and that would mean losing the ability to distinguish among things, since we use spatial resources to identify things as particulars: "Understanding consciousness as it really is would require us to jettison the spatial skeleton of our thought—leaving us with no propositions and so nothing with which to think" (Rowlands 2010, p. 341).

However, regardless of whether consciousness is indeed a nonspatial phenomenon (Gundersen 2011), it is far from certain that concept formation in the empirical sciences is always analogical, or analogical to spatial concepts. In Physics,[2] e.g., the concept of *Minkowski Space-Time* as used in the Theory of Relativity, might be seen as a concept which was formed not on the basis of, or, rather, not analogical to anything in our daily experiences (Walter 1999, p. 50). The same might hold for a concept like *Entanglement* in Quantum Mechanics. On the other hand, concepts such as *Energy* and *Entropy* are non-spatial but have analogues in counting and numbering. Nevertheless, these concepts are not (strictly) mathematical. They are —unlike mathematical concepts— not a priori and not independent of perception as McGinn would have it (McGinn 1989, p. 361). It is, incidentally, interesting to note here that McGinn-confronted with the problem of understanding how something essentially non-spatial can be produced by something essentially spatial, speculates that such an understanding would require us to understand that space has an hitherto hidden nature. We would, then, need a new conception of space if we are to solve the problem of consciousness: "That which we refer to when we use the word 'space' has a nature that is quite different from how we standardly conceive it to be; so different, indeed, that it is capable of 'containing' the non-spatial (as we now conceive it) phenomenon of consciousness" (McGinn 2004, p. 105). It is, again, because of our in-built sensitivity to spatial concepts that, according to McGinn, there is something about physical reality which we are debarred from understanding: "The brain must have aspects that are not represented in our current physical world-view, aspects we deeply do not understand, in addition to all those neurons and electro-chemical processes. There is, on this view, a radical incompleteness in our view of reality, including physical reality" (McGinn 2004, p. 104). Given that new (non-analogical) conceptions of space or, space-time, are indeed possible, McGinn's speculation about the nature of and limits on (human) concept formation as analogical, and its consequences for cognitive closure, cannot be totally convincing. But there is also more. When it comes to identifying the property P, the causal basis of consciousness, McGinn explicitly holds the view that that property is *noumenal* to us (that we are cognitively closed to it) and that it cannot be explained in *any* science *we* can come up with, even though

---

[2]Prof. Christian Maes, Institute for Theoretical Physics KU Leuven, personal communication.

it is true that there is such a property (8c-d): "It is surely possible that we could never arise at a grasp of P; there is, as I said, no guarantee that our cognitive powers permit the solution of every problem we can recognize" (McGinn 1989, p. 353).

McGinn is appealing to the idea of a possible science. However, it is nearly impossible to say what corresponds to such an idea, unless it is defined from within that we call science. In 2.3 it was, e.g., already mentioned that- according to Hume- the relation of cause and effect is essential for existence-claims. At the same time, given that the relation of cause and effect is also experiential, it implies a theory of possible experiences. Both claims, then, are modal claims about what constitutes a possible science and a possible experience. Moreover, in view of McGinn's naturalism, (8d) expresses not just a contingent truth but a necessary truth. This is probably also why (cfr. §2.2) McGinn, when he speculates about what kind of mind could conceive of a solution to the mind-body problem, he makes an analogy with knowledge of a priori (and hence necessarily true) propositions; he says that such a mind "would conceive of the psychophysical link in totally a priori terms" (McGinn 1989, p. 361). *However, in order for (8d) to be necessarily true, it should be true in all possible worlds that are accessible from our world* (Gamut 1991, p. 22). What McGinn is describing in (8d) is a proposition which, in order for it to be true, it should be true in a world that is in principle inaccessible from our world, since that is what the proposition says, which makes it difficult to believe that (8d) is a proposition that is necessarily true. It would seem, then, that (8d) cannot be necessarily true and might, therefore, be false. So, if (8d) is necessarily true, it might be false, which is a contradiction. Therefore (8d) is not necessarily true, McGinn's speculations notwithstanding. Alternatively, one might also say that naturalism and realism are false, since the combination of naturalism and realism forces McGinn to say that (8d) is necessarily true. Borrowing again Dummett's terminology we might say that what McGinn is trying to establish is that because of his semantic realism some propositions have what we could call *absolutely experience-transcending truth conditions*. The problem is the combination of naturalism, effective decidability, absolutely experience-transcending truth-conditions and the concept of cognitive closure. From our discussion it appears that such a combination is incoherent.

4 Conclusion: Mysterianism Exorcised and Gödelized

We have seen how McGinn's mysterianism in its use of analogies w.r.t. concept formation, compositionality, and the a priori in combination with his Transcendental Naturalism and a realist semantics seem to force him —on pain of incoherence— into giving up naturalism or realism (or both). Our criticism, has therefore been mainly semiotic, since it has concentrated on the use of analogy and the semantics of language. McGinn's proposals are, then, not just based on epistemological but also on more general (semiotic) presuppositions. Many other kinds of objections against his proposals are possible (Rowlands 2010, p. 341ff), but it would seem that the semiotic ones have up to now not been highlighted enough.

Even though mysterianism as defended by McGinn is questionable to say the least it is not necessarily completely misdirected. When McGinn claims that his arguments show that "there is, on this view, a radical incompleteness in our view of reality, including physical reality" (McGinn 2004, p. 104), he might have a point. There does indeed seem to be an explanatory gap in our understanding of consciousness as McGinn rightly indicates and it has to do with the fact that conscious states are phenomenal states. In other words, conscious states are qualitative in character and that is also why the problem of consciousness is sometimes identified with the problem of qualia (cfr. §2.1, §2.3.3). On the face of it and up to a certain extent, this explanatory gap is a consequence of an asymmetry between the third-and first-person perspective: "Due to this asymmetry, we have to expect such a 'phenomenal difference' even if we observe one and the same process from either perspective, and we can explain why this is so" (Pauen 1998, p. 204). But maybe something else is also to be considered. David Chalmers notes that there are certain structural analogies between the physical and the phenomenal aspects of information: "There is a direct isomorphism between certain physically embodied information spaces and certain phenomenal (or experiential) information spaces ... That is, we can find the same abstract information space embedded in physical processing and in conscious experience" (Chalmers 1995, p. 216). And further:

> Where there is information, there are *information states* embedded in an *information space*. An information space has a basic structure of difference relations between its elements, characterizing the ways in which different elements in a space are similar or different, possibly in complex ways. An information space is an abstract object, but following Shannon we can see information as *physically embodied* when there is a space of distinct physical states, the differences between which can be transmitted down some causal pathway. The states that are transmitted can be seen as themselves consti-

tuting an information space. To borrow a phrase from Bateson (1972), physical information is a *difference that makes a difference.* (Chalmers 1995, p. 216)

This shifts the focus from finding McGinn's property P (1a-e) to the kind of information-processing this property is supposed to perform. What Chalmers is claiming is that phenomenal properties (qualia) constitute the internal aspect of information (Chalmers 1995, p. 217). This is what he calls the Double-Aspect Theory of information, and it is one of a number of psychophysical principles, principles connecting the properties of physical processes to the properties of experience: "We can think of these principles as encapsulating the way in which experience arises from the physical" (Chalmers 1995, p. 212). It is here that, in our view, Gödel's incompleteness theorems may have a role to play. Indeed, where information is involved questions about the computability of this information almost immediately arise. Assuming that there is a property of the brain that is responsible for consciousness (e.g a property of the brain circuits) and if it so happens that brain circuits can only be modeled by means of a logic that is strong enough to express elementary arithmetic, which —if possible— is highly probable, then by Gödel's (first) incompleteness theorem there will be truths about those circuits that can not be proved in any theory in the neurosciences, which would constitute an obstacle to the reduction of human conscious states to neurophysiological states because there is no complete description of the neurophysiological states (Buechner 2010, p. 32). Penrose (1994), like McGinn, has indicated that one of the reasons why we do not understand consciousness yet has to do with a kind of epistemological incompleteness in our view of reality. He thinks that "appropriate physical action of the brain evokes awareness, but this physical action cannot even be properly simulated computationally" (Penrose 1994, p. 12). 'Awareness' constitutes what Penrose calls the 'passive aspect' of consciousness. He discusses mathematical creativity and uses Gödel's incompleteness theorems to show how human beings can "see" the truth of certain mathematical propositions even though within an axiomatic system these propositions are undecidable, i.e. neither the propositions themselves nor their negations can be proven, which is some kind of non-algorithmic insight. This allows him to conclude that thinking (and by extension consciousness) is non-algorithmic and, though it can be explained scientifically, the kind of theory that will be needed will have to be non-computational, arguing that a new kind of quantum physics has to be developed whose (non-computable) laws explain mental processes and consciousness : "The final conclusion of all this is rather alarming. For it suggests that we must seek a non-computable physical theory that reaches beyond every computable level of Oracle machines (and perhaps beyond)" (Penrose 1994, p. 381). One is reminded here of McGinn's earlier remark (§3.3) that there is a radical incompleteness in our view of physical reality. Perhaps Penrose's speculation is true, perhaps it is false, but it shows that it may indeed be the case that incompleteness is part of the reason why we do not yet fully understand consciousness. What Penrose is in fact suggesting is that we look for a non-computational basis for computational thinking and for consciousness. Now, admittedly, this presents a kind of mysterianism of its own. We do not yet know what such a physics would be like, but, unlike McGinn, the proposal at least does not make an appeal to a science that is completely out of reach for us and the idea is not necessarily incoherent such as is the case with McGinn's proposal. But even if Penrose's proposal for a new physics does not ring true, there is an analogy we can draw which may elucidate partly the nature of conscious processes, viz. their subjectivity or 'phenomenality'. Gödel's incompleteness theorems are about logical incompleteness. If Pauen (1998) is right about the difference between the first-and third person point of view in accounting for the explanatory gap and if Chalmers (1995) is right in saying that phenomenal properties constitute the internal aspect of the information processed in conscious states, then a metalogical property such as incompleteness which is definitely a property of (complex) formalized mathematical systems is also a property of the information the system contains. If a system is logically incomplete it means that it has non-isomorphic models, also called non-standard models, which is for instance the case with the models of formalized arithmetic in Gödel's theorems. This comes down to the fact that the information in the axioms of the theory is in a sense insufficient to uniquely determine what they are "about". There is, therefore, a 'tension' between the 'inside' (information) of the theory and the 'outside' of what it is about (the 'world'). Assuming that the information processed in conscious states also exhibits this kind of incompleteness-which is not unreasonable to suppose since, as Penrose would have it, human consciousness is able to process complex mathematical systems amongst other things, this might- somewhat analogously - phenomenally be experienced as a 'felt' distinction between the inside of the system (the conscious mind) and the outside world with which the system interacts and therefore create the feeling of subjectivity. In this sense the first-person or subjective point of view has a logical analogon in the incompleteness of the information processed (see also Vergauwen 2012). This does not exclude that conscious processes are identified with brain processes, but it does suggest that the quest for the nature of consciousness should be pursued on a higher level of information processing even though certain properties of the brain are causally responsible for consciousness. Such a view may avoid the difficulties and pitfalls that go with McGinn's mysterianism without necessarily leading to a reductionist naturalism.

Be that as it may, paraphrasing Winston Churchill's famous Russia-quote, it seems fair to say that if *Consciousness is a Riddle wrapped in a Mystery inside an Enigma* we have probably gone past the Mystery by now, which leaves us to tackle the Riddle and the Enigma. But optimism is a moral duty.

References

Borge, S. (2006). "Defending the Martian Argument". In: *Disputatio* 1, pp. 337–345.
Buechner, J. (2010). "Are the Gödel Incompleteness Theorems Limitative Results for the Neurosciences?" In: *Journal of Biological Physiscs* 36, pp. 23–44.
Chalmers, D. (1995). "Facing up the Problem of Consciousness". In: *Journal of Consciousness Studies* 2.3, pp. 200–219.
Coolidge, F. L., K. A. Overmann, and T. Wynn (2011). "Recursion: What is it, who has it, and how did it evolve?" In: *Wiley Interdisciplinary Reviews: Cognitive Science* 2.5, pp. 547–554.
Dennett, D. (1991). *Consciousness Explained*. London: Penguin Books.
Devitt, M. and K. Sterelny (1989). "Linguistics: What's wrong with 'The Right View'?" In: *Philosophical Perspectives. Philosophy of Mind and Action Theory* 3, pp. 498–531.
Dummett, M. (1976). "What is A Theory of Meaning? (II)". In: *Truth and Meaning: Essays in Semantics*. Ed. by G. Evans and J. McDowell. Oxford University Press, pp. 67–137.
— (1982). "Realism". In: *Synthese* 52, pp. 55–112.
Flanagan, O. (1992). *Consciousness Reconsidered*. Cambridge, MA: MIT Press.
Gamut, L. T. F. (1991). *Logic, Language, and Meaning: Volume 2, Intensional Logic and Logical Grammar*. Chicago and London: University of Chicago Press.
Grayling, A. C. (1982). *An Introduction to Philosophical Logic*. Sussex: Harvester Press.
Gundersen, S. (2011). "Is Consciousness a Nonspatial Phenomenon?" In: *Kritike* 5.1, pp. 91–98.
Haack, S. (1974). *Deviant Logic*. Cambridge: Cambridge University Press.

Hauser, M., N. Chomsky, and W. Fitch (2002). "The Faculty of Language: What Is It, Who Has It, and How Did It Evolve?" In: *Science* 298.22, pp. 1569–1579.
Hintikka, J. (1975). "A Counterexample to Tarski-Type Truth Conditions as applied to Natural Languages". In: *Philosophia* 5, pp. 207–212.
— (1984). "A Hundred Years Later: The Rise and Fall of Frege's Influence in Language Theory". In: *Synthese* 59, pp. 27–49.
— (1996). *The Principles of Mathematics Revisited*. Cambridge: Cambridge University Press.
Hintikka, J. and G. Sandu (1997). "Game-Theoretical Semantics". In: *Handbook of logic and Language*. Ed. by J. Van Benthem and A. Ter Meulen. Amsterdam and Cambridge: Elsevier and MIT Press, pp. 361–410.
Hume, D. (1777). *Enquiries concerning the Human Understanding and concerning the Principles of Morals*. Ed. by L. A. Selby-Bigge. Oxford: Clarendon Press.
Janssen, T. M. V. (1997). "Compositionality". In: *Handbook of logic and Language*. Ed. by J. Van Benthem and A. Ter Meulen. Amsterdam and Cambridge: Elsevier and MIT Press, pp. 417–473.
Kirkham, R. L. (1997). *Theories of Truth*. Cambridge, MA: MIT Press.
Laurence, S. (2003). "Is Linguistics a Branch of Psychology?" In: *Epistemology of Language*. Ed. by A. Barber. Oxford University Press, pp. 69–106.
Levine, J. (1983). "Materialism and Qualia: The Explanatory Gap". In: *Pacific Philosophical Quarterly* 64, pp. 354–361.
McGinn, C. (1979). "An A Priori Argument for Realism". In: *Journal of Philosophy* 76.3, pp. 113–133.
— (1982). "Realist Semantics and Content Ascription". In: *Synthese* 52, pp. 113–124.
— (1989). "Can We Solve the Mind-Body Problem?" In: *Mind* 98.391, pp. 349–366.
— (1993). *Problems in Philosophy: The Limits of Enquiry*. Blackwell USA.
— (2004). *Consciousness and Its Objects*. Oxford University Press.
Nagel, T. (1972). "What is It like to be a Bat ?" In: *Philosophical Review* 82, pp. 435–456.
Pagin, P. and D. Westertahl (2010). "Compositionality II: Arguments and Problems". In: *Philosophy Compass* 5.3, pp. 265–282.
Pauen, M. (1998). "Is there an Empirical Answer to the Explanatory Gap?" In: *Consciousness and Cognition* 7, pp. 202–205.

Penrose, R. (1994). *Shadows of The Mind: A Search for the Missing Science of Consciousness*. Oxford: Oxford University Press.

Putnam, H. (1980). "Models and Reality". In: *The Journal of Symbolic Logic* 45, pp. 464–482.

Rowlands, M. (2010). "Mysterianism". In: *The Blackwell Companion to Consciousness*. Ed. by M. Velmans and S. Schneider. Blackwell Publishing, pp. 335–345.

Searle, J. (1997). *The Mystery of Consciousness*. London: Granta Books.

Sher, G. (2001). "Truth, Logical Structure, and Compositionality". In: *Synthese* 126, pp. 195–219.

Van Bendegem, J. P. (2000). "Analogy and Metaphor as essential Tools for the working Mathematician". In: *Metaphor and Analogy in the Sciences*. Ed. by F. Hallyn. Dordrecht: Kluwer Academic Publishers, pp. 105–123.

Vergauwen, R. (1992). *A Metalogical Theory of Reference*. Lanham, MD: University Press of America.

— (2012). "Language as Embodied Information". In: *Logique et Analyse* 218, pp. 129–146.

Walter, S. (1999). "Minkowski, Mathematicians, and the Mathematical Theory of Relativity". In: *The Expanding Worlds of General Relativity*. Ed. by H. Goenner et al. Vol. 7. Einstein Studies. Boston and Basel: Birkhäuser, pp. 45–86.

www.ingramcontent.com/pod-product-compliance
Lightning Source LLC
Chambersburg PA
CBHW070052200426
43197CB00051B/1975